Current Topics in
Developmental Biology

Volume 26
Cytoskeleton
in Development

Series Editor

Roger A. Pedersen
Laboratory of Radiobiology
and Environmental Health
University of California
San Francisco, CA 94143

Editorial Board

John C. Gerhart
University of California, Berkeley

Peter Gruss
Max-Planck-Institute
of Biophysical Chemistry, Göttingen-Nikolausberg

Philip Ingham
Imperial Cancer Research Fund, Oxford

Story C. Landis
Case Western Reserve University

David R. McClay
Duke University

Gerald Schatten
University of Wisconsin, Madison

Virginia Walbot
Stanford University

Mitsuki Yoneda
Kyoto University

Founding Editors

A. A. Moscona
Alberto Monroy

Current Topics in Developmental Biology

Volume 26
Cytoskeleton in Development

Edited by

Elaine L. Bearer
Division of Biology and Medicine
Brown University
Providence, Rhode Island

Academic Press, Inc.
Harcourt Brace Jovanovich, Publishers
San Diego   New York   Boston   London   Sydney   Tokyo   Toronto

COLORADO COLLEGE LIBRARY
COLORADO SPRINGS
COLORADO
WITHDRAWN

Front cover photograph: Time sequence showing normal development of *Strongylocentrotus purpuratus* embryos at 15°C. See page 72 for details. Adapted from Davidson *et al.* (1982, Fig. 1); copyright 1982 by the AAAS.

.

This book is printed on acid-free paper. ∞

Copyright © 1992 by ACADEMIC PRESS, INC.

All Rights Reserved.

No part of this publication may be reproduced or transmitted in any form or by any means, electronic or mechanical, including photocopy, recording, or any information storage and retrieval system, without permission in writing from the publisher.

Academic Press, Inc.
1250 Sixth Avenue, San Diego, California 92101

United Kingdom Edition published by
Academic Press Limited
24–28 Oval Road, London NW1 7DX

Library of Congress Catalog Number: 66-28604

International Standard Book Number: 0-12-153126-0

PRINTED IN THE UNITED STATES OF AMERICA
92 93 94 95 96 97   BB   9 8 7 6 5 4 3 2 1

Contents

Introduction
Elaine L. Bearer

1

Actin Organization in the Sea Urchin Egg Cortex
Annamma Spudich

2

Localization of *bicoid* Message during *Drosophila* Oogenesis
Edwin C. Stephenson and Nancy J. Pokrywka

3

Actin and Actin-Associated Proteins in *Xenopus* Eggs and Early Embryos: Contribution to Cytoarchitecture and Gastrulation
Elaine L. Bearer

4

Microtubules and Cytoplasmic Reorganization in the Frog Egg
Evelyn Houliston and Richard P. Elinson

5

Microtubule Motors in the Early Sea Urchin Embryo
Brent D. Wright and Jonathan M. Scholey

Contents

6

Assembly of the Intestinal Brush Border Cytoskeleton
Matthew B. Heintzelman and Mark S. Mooseker

7

Development of the Chicken Intestinal Epithelium
Salim N. Mamajiwalla, Karl R. Fath, and David R. Burgess

8

Developmental Regulation of Sarcomeric Gene Expression
Charles P. Ordahl

Contributors

Numbers in parentheses indicate the pages on which the authors' contributions begin.

Elaine L. Bearer Division of Biology and Medicine, Brown University, Providence, Rhode Island 02912 (1, 35)

David R. Burgess Department of Biological Sciences, University of Pittsburgh, Pittsburgh, Pennsylvania 15260 (123)

Richard P. Elinson Department of Zoology, University of Toronto, Toronto, Ontario M5S 1A1, Canada (53)

Karl R. Fath Department of Biological Sciences, University of Pittsburgh, Pittsburgh, Pennsylvania 15260 (123)

Matthew B. Heintzelman Department of Biology, Yale University, New Haven, Connecticut 06511 (93)

Evelyn Houliston Department of Zoology, University of Toronto, Toronto, Ontario M5S 1A1, Canada (53)

Salim N. Mamajiwalla Department of Biological Sciences, University of Pittsburgh, Pittsburgh, Pennsylvania 15260 (123)

Mark S. Mooseker Department of Biology, Yale University, New Haven, Connecticut 06511 (93)

Charles P. Ordahl Department of Anatomy, University of California, San Francisco, San Francisco, California 94143 (145)

Nancy J. Pokrywka Department of Biology, University of Rochester, Rochester, New York 14627 (23)

Jonathan M. Scholey Department of Zoology, University of California, Davis, Davis, California 95616 and Department of Cellular and Structural Biology, University of Colorado Health Science Center, Denver, Colorado 80262 (71)

Annamma Spudich Department of Cell Biology, School of Medicine, Stanford University, Stanford, California 94305 (9)

Edwin C. Stephenson Department of Biology, University of Rochester, Rochester, New York 14627 (23)

Brent D. Wright Department of Zoology, University of California, Davis, Davis, California 95616 and Department of Cellular and Structural Biology, University of Colorado Health Science Center, Denver, Colorado 80262 (71)

Preface

A major question in developmental biology is how the radially symmetric egg transforms into the complex multicellular embryo. In some species the initial coordinates for this transformation are already determined by the localized accumulation of positional cues which are distributed asymmetrically in the highly structured cytoplasm of the fertilized egg. This cytoplasmically localized information is initially localized during oogenesis, redistributed during fertilization, and then differentially segregated into daughter cells during early cleavage stages. In some as yet poorly defined way, that information subsequently specifies how the individual cells that inherit it behave during gastrulation so that the radially symmetric blastula is transformed into a trilaminar embryo.

Both the cytoplasmic structure of the egg and the cellular movements of gastrulation depend upon the cytoskeleton, which includes the filamentous proteins actin, tubulin, and intermediate filaments. This book presents essays on the structure and molecular composition of actin and tubulin and their contribution to the early events of fertilization and gastrulation. The book concludes with essays exploring the temporally regulated expression of proteins necessary for structural differentiation occurring later in development, such as formation of intestinal microvilli or the contractile machinery of skeletal muscle.

This volume continues a tradition in *Current Topics in Developmental Biology* of devoting special issues to topical subjects that are important for development. The cytoskeleton is a prime example of how structure and function are inextricably related in developing systems. Like a previous volume, *Growth Factors and Development* (1990), this volume explores in depth a subject that is represented by several experimental systems. This reflects our increasing awareness of the universality of developmental mechanisms. Thus, *Cytoskeleton in Development* complements the eclectic volumes in this series, which call attention to the diversity of developmental strategies represented in the formation of various plant and animal body plans.

Elaine L. Bearer
Guest Editor

Roger A. Pedersen
Series Editor

Introduction

Elaine L. Bearer
Division of Biology and Medicine
Brown University
Providence, Rhode Island 02912

I. Positional Information and the Cytoskeleton

In the past few years, Christiane Nusslein-Volhard and co-workers have revealed the molecular nature of this positional information in *Drosophila* embryos (reviewed in Nusslein-Volhard and Roth, 1989). To do this, they made EMS-induced point mutations in *Drosophila* and looked for homozygous mutant females that failed to produce viable embryos but were otherwise normal. In this way, they were able to identify a set of genes that were necessary only in the mother and only for the production of a normal embryo. These mutants were termed "maternal effect." The mutations that affected the embryos fell into four groups: those that altered the dorsal–ventral body axis (Anderson and Nusslein-Volhard, 1984); those that altered the anterior structures; those that altered posterior-abdomenal structures (Schupbach and Wieschaus, 1986); and those that affected terminal structures at both ends of the embryo (Nusslein-Volhard *et al.*, 1987). Similar screens in other laboratories have produced additional mutations that are also in genes necessary for specification of the body axis (Stephenson and Mahowald, 1987), while other genes were found because of their homology to homeobox proteins (Mlodzik *et al.*, 1985).

The genes for several of these mutations have been cloned, and the pattern of expression and distribution of the mRNA and the protein have been examined. The localized distribution of mRNA appears to be necessary for the proper activity of *bicoid*, required for formation of the head (Berleth *et al.*, 1988). In contrast, the posterior-group genes appear to depend upon localized deposition of the protein product to the posterior region of the egg (Lehman and Nusslein-Volhard, 1986). It is not yet clear what asymmetric event specifies the dorsal–ventral or the terminal-group genes. These two groups include genes that encode putative transmembrane receptors, homologous to receptors found in other trans-membrane signaling pathways (examples include the protein product of Toll, in the dorsal–ventral group (Hashimoto *et al.*, 1988); and the torso gene product in the terminal group (Sprenger *et al.*, 1989). Thus, the establishment of either of these embryonic coordinates may depend upon a signal transduction pathway such as that proposed to be the basis of induction. Whether these receptors are asymmetrically activated, and what molecules activate them, are not yet known.

It seems likely that the cytoskeleton should be involved in the active transport of the molecules that connote positional information to their appropriate place in the cytoplasm, and in the stabilization of that information. However, in *Drosophila* very little is known about the contribution of the cytoskeleton to the asymmetries of the egg (reviewed in Bearer, 1991). *Drosophila* eggs are fertilized internally by the entrance of the spermatozoan through the micropile, an opening in the egg investitures whose position is determined by the female. This means that the site of sperm entry, which influences the subsequent polarity of the egg in other organisms, is not necessarily the determining factor in the establishment of the embryonic axis in fly. Thus, some of the cytoplasmic re-distributions that accompany fertilization in other creatures would not necessarily occur in fly. However, because internal fertilization makes observation difficult in fly, these events have yet to be described.

The anterior or head region of the *Drosophila* embryo is established early in oogenesis and depends upon the localized accumulation of the *bicoid* messenger RNA (Frigerio *et al.*, 1986). In this book, Edwin Stephenson has contributed a chapter discussing the role of microtubules and the *swallow* gene product in establishing and maintaining that localization. Further insights into the process of localizing positional cues will require more detailed examination of microtubules and actin in eggs and ovaries during the time when the oocyte cytoplasm is forming and becoming structured, as well as direct observation of the actin cytoskeleton during egg activation and fertilization. These studies are now in progress but are not yet at a point to be presented in this review volume.

Much more is known about the role of the cytoskeleton in establishing embryonic coordinates in other species. Both actin and tubulin have been implicated in the reorganization of the cytoplasm and the localization of positional information. In the Nematode (*Caenorhabitis elegans*) and the Ascidian (*Styela*), contraction of cortical actin filaments results in relocation of cytoplasmic components. This contraction of the cortex which follows fertilization has been described in recent reviews (Strome and Hill, 1988; Jeffery and Swalla, 1990), and thus is not represented here.

In amphibians, the rotation of the cytoplasm which describes the anterior–posterior axis depends on microtubules (Manes *et al.*, 1978). In this book, the chapter contributed by Houliston and Elinson reviews the role of microtubules and microtubule motors in the rotation of the amphibian egg cytoplasm, which defines the dorsal lip of the blastopore and consequently the anterior–posterior axis of the embryo.

While the contractions and rotations of the egg that follow fertilization depend upon the cytoskeleton, the molecules that regulate these cytoskeletal behaviors are not yet defined. For this reason, the majority of chapters in this book review what is known about the cytoskeletal machinery of eggs and early embryos, even though the relationship of these molecules to either positional cue localization or to the specific cytoskeletal behaviors that mediate that localization is not known. This is done in the belief that this information, presented in the context of the question of egg asymmetries and subsequent embryonic structure that arises from it, will lead to insights and ideas that will prompt more direct exploration of the links between the cytoskeleton and positional information. In this book, Spudich presents a chapter describing the changes in actin filaments in the cortex of the sea urchin egg during fertilization. This chapter describes actin-binding proteins purified from these eggs that are potentially regulators of actin dynamics. In their chapter, Wright and Scholey present detailed information on the biochemical behavior of microtubule motors isolated from sea urchin eggs. Such motors are likely to mediate transport of positional information from one part of the cytoplasm to another, or to be involved in cytoplasmic rotations.

In these systems, much more is known about the cytoskeleton than about the molecular nature of the positional information. However, ultimately it may be possible to juxtapose the knowledge about positional information gleaned from fly with cytoskeletal behaviors such as those described in this volume. It is becoming increasingly apparent that developmental genes originally identified in fly have homologs that perform analogous functions in other species. For example, the *Drosophila* maternal effect gene *dorsal* has partial sequence homology to the rel family of oncogenes (Steward, 1987). A *Xenopus* rel/dorsal-like protein is present in embryos (Bearer, manuscript in preparation), and a transcript homologous to rel has been identified in *Xenopus* ovaries and early embryos (Kao and Hopgood, 1991). The zygotic segmentation gene, *engrailed,* has been found in both mouse (reviewed in Joyner and Martin, 1987) and frog (Brivanlou and Harland, 1989), but is expressed too late to perform an analogous function. The zygotic gene *twist,* identified in *Drosophila* as part of the dorsal–ventral axis, is also found in frog (Hopgood *et al.,* 1989). Thus, homologs of the *Drosophila* gene products involved in specifying embryonic coordinates may be observed in other species in a context where cytoskeletal behavior is more easily observed and already better defined. These organisms should ultimately prove to be an accessible testing ground for paradigms of the mechanisms involved in positional cue localization.

Conversely, the cytoskeletal insights gained in other species might provide clues as to what sorts of events to look for in fly embryos that could be involved in the localization of identified positional information. For example, since actin filament-based contractions are part of redistribution of posterior cues in nematodes and ascidia, it would be interesting to observe whether such contradictions also follow fertilization in the fly, and if they play a role in delivering such posterior cues as the gene products of any of the posterior-group maternal effect genes, such as *vasa, oskar,* or *nanos.*

II. Gastrulation

The cytoskeleton must play a major role in determining the behavior of individual cells during the movements of gastrulation. These movements are a response to the information initially specified by the positional cues laid out in the egg cytoplasm and the zygotic genes that result from that specification. While actin has been extensively investigated in migrating fibroblasts in culture taken from late (day 9–10) chick embryo body wall, very little is yet known about the specific behavior of actin in the motile cells of the gastrulating embryo. Indeed, only recently has it been possible to obtain in-depth information concerning the individual behavior of these cells in embryos during gastrulation using video-enhanced microscopy (Keller, 1984; Hardin and Keller, 1988; Fink and Trinkaus, 1988). A video tape is now available for purchase containing time-lapse microscopy of 12 different embryos during development (Fink, 1991). It now appears that the behavior of cells in culture may be quite different from that of the motile cell in the embryo, in that the cells in embryos do not elaborate lamellopodia but rather extend long filopodia. This morphological different must be reflected in the molecular mechanisms that are responsible for cell migration and particular to the mechanisms of actin filament behavior.

It seems likely that one way in which positional information directs the process of gastrulation is through the regulation of genes encoding molecules that themselves regulate cytoskeletal changes upon which changes in cell shape and cell migration depend. Thus, a migrating cell of the mesoderm would be expected to express migration-specific genes, whose transcription might be induced by the *dorsal* gene product. Conversely, an epithelial cell might express cytoskeletal genes required for the formation of cell–cell junctions and cell spreading. The patterns of expression of particular cytoskeletal proteins in these different embryonic tissues could reveal something about their role in these different cellular behaviors. In the chapter by Bearer, the expression pattern in the frog embryo of a novel actin-associated protein is described, and placed in the context of actin dynamics known to occur in frog embryos and in the context of the cellular movements and shape changes that permit gastrulation.

III. Formation of Tissue-Specific Cytoskeletal Structures

Actin filaments and microtubules are ubiquitous structures in eukaryotic cells. In addition to their involvement in basic cellular processes common to all cells, they are also used as building blocks for tissue-specific structures such as microvilli, contractile elements of muscle, and cilia and flagella. The two structures presented in this book are both actin based—microvilli and the contractile machinery of muscle. The pattern of expression of the proteins of the intestinal microvillus is described by Heintzelman and Mooseker, while the expression of message is further explored in the chapter by Mamajiwalla, Fath, and Burgess. The pattern of expression of the contractile machinery is at a more advanced stage. In this system, the promoter–enhancer regions that regulate expression of muscle-specific genes are known, some of the transcription factors that interact with these regulatory sequences are known, and many muscle-specific proteins have been purified, sequenced, and their expression patterns carefully delineated. This information is presented in a detailed, lucid chapter by Ordahl.

In order for a cell to elaborate complex structures such as microvilli or contractile machinery, it must express the other required components in a sequential and organized pattern as part of its differentiation program. Some of these components, such as brush border myosin I, appear to be unique for that particular structure and are not expressed at appreciable levels in cells that do not elaborate microvilli, while others, such as tropomyosin, are involved in many other actin-based structures.

These last three chapters may seem far afield from the events of early embryogenesis. They are included here for two reasons. First, some of the cytoskeletal proteins involved in ooplasmic segregation, localization of positional information, and cytoplasmic reorganization following fertilization must be specific to the egg, and thus tissue specific. The expression pattern of these ovary-specific genes may help elucidate these embryonic events. It is hoped that the methods and approaches used to elucidate tissue-specific expression of components necessary for other unique structures may prove inspirational to exploration of the ovary.

Second, during gastrulation populations of cells behave differently. This difference in behaviors, such as motility or contractions, could be due to a different response to an inductive signal that results in posttranslational modification of cytoplasmic proteins that then directly mediate the cellular behavior. Alternatively, it is conceivable that part of the developmental program that establishes the fate map involves the differential expression of those genes required for shape changes or cellular movements in cells that are programmed to move. Thus, understanding the mechanisms that specify migration in embryos may produce information regarding the specification of the fate map and future differentiation.

This set of essays was compiled selfishly—because I wish to learn more about the contribution of the cytoskeleton to crucial embryonic events and believed that juxtaposition of information gained from different species and with different approaches might lead to the identification of paradigms. Furthermore, I hoped that reviewing what is known would reveal the next best experiment that would resolve discrepancies, fill in critical gaps, and identify future directions.

References

Anderson, K. V., and Nusslein-Volhard, C. (1984). *Dorsal* group genes of *Drosophila*. In "Gametogenesis and the Early Embryo" (J. Gall, ed.), pp. 177–194. Alan Liss, New York.

Bearer, E. L. (1991). Actin in the *Drosophila* embryo: Is there a relationship to developmental cue localization? *BioEssays* **13**, 199–204.

Berleth, T., Burri, M., Thoma, G., Bopp, D., Richstein, S., Frigerio, G., Noll, M., and Nusslein-Volhard, C. (1988). The role of localization of *bicoid* RNA in organizing the anterior pattern of the *Drosophila* embryo. *EMBO J.* **7**, 1749–1756.

Brivanlou, A. H., and Harland, R. M. (1989). Expression of an engrailed-related protein is induced in the anterior neural ectoderm of early *Xenopus* embryos. *Development* **106**, 611–617.

Fink, R.D., and Trinkaus, J. P. (1988). Fundulus deep cells: Directional migration in response to epithelial wounding. *Dev. Biol.* **129**, 179–190.

Fink, R.D., editor (1991). "A dozen eggs: Time lapse microscopy of normal development." Society for Dev. Biol and Sinauer Assoc., Sunderland, Mass. A 45-minute video tape.

Frigerio, G., Burri, M., Bopp, D., Baumgartner, S. and Noll, M. (1986). Structure of the segmentation gene paired and the *Drosophila* PRD gene set as part of a gene network. *Cell* **47**, 735–739.

Hardin, J., and Keller, R. (1988). The behavior and function of bottle cells during gastrulation in *Xenopus laevis*. *Development* **103**, 211–230.

Hashimoto, C., Hudson, K. L., and Anderson, K. V. (1988). The Toll gene of *Drosophila*, required for dorsal–ventral polarity, appears to encode a transmembrane protein. *Cell* **52**, 269–279.

Hopgood, N. D., Pluck, A., and Gurdon, J. B. (1989). A *Xenopus* mRNA related to *Drosophila* twist is expressed in response to induction in the mesoderm and the neural crest. *Cell* **59**, 863–903.

Jeffery, W. R., and Swalla, B. J. (1990). The myoplasm of ascidian eggs: A localized cytoskeletal domain with multiple roles in embryonic development. *Semin. Cell Biol.* **1**, 373–381.

Joyner, A. L., and Martin, G. R. (1987). En-1 and En-2, two mouse genes with sequence homology to the *Drosophila* engrailed gene: Expression during embryogenesis. *Genes Dev.* **1**, 124–138.

Kao, K. R., and Hopgood, N. D. (1991). Expression of mRNA related to c-rel and *dorsal* in early *Xenopus laevis* embryos. *Proc. Natl. Acad. Sci. U.S.A.* **88**, 2697–2701.

Keller, R. E. (1984). The cellular basis of gastrulation in *Xenopus laevis*: Active, post-involution convergence and extension by mediolateral interdigitation. *Am. Zool.* **24**, 589–603.

Lehman, R., and Nusslein-Volhard, C. (1986). Abdomenal segmentation, pole cell formation, and embryonic polarity require the localized activity of *oskar*, a maternal gene in *Drosophila*. *Cell* **47**, 141–152.

Manes, M. E., Elinson, R. D., and Barbieri, F. D. (1978). Formation of the amphibian grey crescent: Effects of colchicine and cytochalasin B. *Roux's Arch. Dev. Biol.* **185**, 99–104.

Mlodzik, M., Fjose, A., and Gehring, W. (1985). Isolation of caudal, a *Drosophila* homeo box-containing gene with maternal expression, whose transcripts form a concentration gradient at the pre-blastoderm stage. *EMBO J.* **4**(11), 2961–2969.

Nusslein-Volhard, C., Frohnhofer, H. G., and Lehman, R. (1987). Determination of anteroposterior polarity in *Drosophila*. *Science* **238**, 1675–1681.

Nusslein-Volhard, C., and Roth, S. (1989). Axis determination in insect embryos. *Ciba Found. Symp.* **144**, 37–55. Discussion 55-64, 92-98.

Schupbach, T., and Wieschaus, E. (1986). Maternal-effect mutations altering the anterior-posterior pattern of the *Drosophila* embryo. *Roux's Arch. Dev. Biol.* **195**, 302–317.

Sprenger, F., Stevens, L. M., and Nusslein-Volhard, C. (1989). The *Drosophila* gene torso encodes a putative receptor tyrosine kinase. *Nature (London)* **338,** 478–483.

Stephenson, E. C., and Mahowald, A. P. (1987). Isolation of *Drosophila* clones encoding maternally restricted RNAs. *Dev. Biol.* **124,** 1–8.

Steward, R. (1987). *Dorsal,* an embryonic polarity gene in *Drosophila,* is homologous to the vertebrate proto-oncogene, c-rel. *Science* **238,** 692–694.

Strome, S., and Hill, D. P. (1988). Early embryogenesis in *C. elegans:* The cytoskeleton and spatial organization of the zygote. *BioEssays* **8,** 145–149.

1

Actin Organization in the Sea Urchin Egg Cortex

Annamma Spudich
Department of Cell Biology
School of Medicine
Stanford University
Stanford, California 94305

I. Introduction

The cortical region of eukaryotic cells is involved in many fundamental cellular activities. The cortex is defined as an ~ 1-μm thick layer of the cytoplasm lying immediately adjacent to the plasma membrane. The cell cortex is a dynamic region involved in many important functions, such as cell shape determination, directed cell movements, traffic in and out of the cell, and adhesion of cells with each other and with a substratum.

The importance of the cell cortex was recognized by early developmental biologists studying fertilization. Those classical studies dealt with understanding the nature and localization of "morphogenetic determinants" thought to be localized in the egg cortex. As early as 1901 Boveri showed that the animal–vegetal axis as well as bilateral symmetry are determined in the unfertilized eggs of many organisms (Davidson, 1986), and the question of how this topological localization is achieved became one of the central questions of early embryology.

In the beginning of this century Conklin (1917) proposed that the localized morphogenetic determinants are held in place by being attached to a "persistent framework" in the cortex of eggs. The persistent cortical framework was proposed to be made up of a "spongioplasm" that was viewed as having elastic and contractile properties. These conclusions by Conklin were the result of experi-

Current Topics in Developmental Biology, Vol. 26
Copyright © 1992 by Academic Press, Inc. All rights of reproduction in any form reserved.

9

ments with centrifuged eggs in which polarity and development of eggs proceeded normally even though cytoplasm was stratified, as indicated by displacement of pigment granules and oil droplets. The nature of the elastic element responsible for tethering the morphogenetic determinants has been the subject of intense study in cell biology in the decades since it was described. Using an "elastimeter" Mitchison and Swann (1955) measured the stiffness of the unfertilized sea urchin egg cortex and concluded that the cortex is a semisolid gel about 1.5 μm thick. Experiments by Hiramoto (1970) using a microinjected iron particle showed the cortex of the sea urchin egg to be gellike and significantly more viscous than the inner cytoplasm.

A number of excellent reviews have covered many aspects of the fertilization process (Hiramoto, 1970; Epel, 1978, 1982; Vacquier, 1981; Shapiro et al., 1981; G. Schatten, 1982). This article focuses on recent findings concerning a novel organizational state of actin in the egg cortex and on how the actin organization might relate to fertilization-induced exocytosis of cortical granules and microvilli elongation.

II. Actin is a Major Component of the Unfertilized Egg Cortex

The biochemical nature of the elastic element in the cortex is starting to be understood. Actin has been shown to be a major component of the echinoderm egg (Mabuchi and Sakai, 1972). Work by Mabuchi and Spudich (1980) showed that about 50% of the actin of the unfertilized egg is high-speed sedimentable, and that more than 80% of the high-speed supernatant actin is monomeric by gel filtration and by the DNase I inhibition assay. Actin purified to homogeneity from high-speed supernatants polymerizes with kinetics similar to that of purified muscle actin, suggesting the presence of inhibitors of polymerization in the high-speed extract.

Actin has been identified as one of the major proteins of the cortex isolated from unfertilized sea urchin eggs (for review, see Vacquier, 1981). About 20% of the total egg actin is found in isolated cortical preparations (Spudich and Spudich, 1979), and as much as 27% of the total protein of the isolated unfertilized egg cortex was estimated to be actin by Vacquier and Moy (1980). Actin can assemble into filaments and filament bundles, which can confer elastic properties to the cortex. As discussed below, modulation of the actin assembly state is seen to accompany many of the functions of the cell cortex. Understanding how these changes are brought about at the molecular level is a fundamental problem of cell biology.

III. Actin is Organized in Two Distinct Assembly States in the Unfertilized Egg Cortex

One assembly state of actin in the unfertilized egg cortex is that of conventional filamentous actin (F-actin). Scanning and transmission electron microscopy (EM) images of the cortex of whole eggs or of isolated fragments of the cortex show tight bundles of short actin filaments in the prefertilization microvilli and longer actin filaments that run along the plane of the plasma membrane (Chandler and Heuser, 1981; Sardet, 1984; Henson and Begg, 1988; Spudich *et al.*, 1988). The microvillar actin is retained in a detergent-insoluble pellet of the cortex (Spudich *et al.*, 1988), suggesting that it is associated with specialized areas of the membrane. The population of sparsely distributed filaments arranged in the plane of the plasma membrane interconnect the microvillar actin bundles (Henson and Begg, 1988; Spudich *et al.*, 1988). The presence of F-actin in the unfertilized egg cortex is also suggested by the work of Cline and Schatten (1986) and Yonemura and Mabuchi (1987), who showed that the fungal toxin phalloidin, a drug known to be specific for F-actin, stains discrete spots in the cortex. The distribution of these phalloidin-staining spots probably corresponds to actin filament cores in the prefertilization microvilli.

The possibility that actin in the unfertilized egg cortex is organized in a second assembly state that is not filamentous was first indicated by biochemical experiments (Spudich and Spudich, 1979). The majority of the actin associated with the unfertilized egg cortex (approximately 80%) is readily solubilized by dialysis against an actin-depolymerizing buffer, leaving about 20% of the actin retained with a high-speed pellet. However, under assembly conditions at actin concentrations well above the critical concentration for assembly, this solubilized actin forms amorphous aggregates rather than conventional filaments. When purified to homogeneity, the high-speed supernatant actin is very similar to purified muscle actin in its ability to form filaments, to form magnesium paracrystals, and to activate myosin ATPase. These observations indicate that the majority of the actin of the unfertilized egg cortex is organized in a form that is readily solubilized and apparently nonfilamentous. Further biochemical work (Spudich *et al.*, 1982) showed that this nonfilamentous actin, which is associated with the cortex, contains ADP as the bound nucleotide and interacts with DNase I with kinetics more similar to that of F-actin than to that of monomeric actin (G-actin). Taken together, the above results are consistent with the presence of some conventional actin filaments in the prefertilization egg cortex and a large amount of actin that, although apparently nonfilamentous, is in some assembled state. This motif may be similar to the organization of actin in the filamentous actomere and the nonfilamentous profilactin material in *Thyone* sperm, as described by Tilney (1976, 1978).

Structural evidence that a population of nonfilamentous actin is present in the

Annamma Spudich

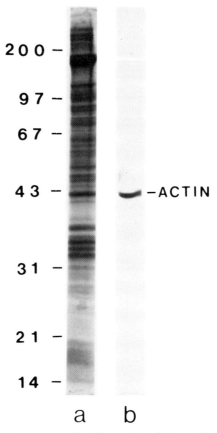

Fig. 1 An SDS gel and the corresponding Western blot of the total lysate of unfertilized *S. purpuratus* eggs. (a) The complex protein pattern of the total egg lysate is shown by a Coomassie-stained SDS gel. The position of different molecular weight standards ($\times$ 10^3) are indicated at the left. (b) The Western blot of the egg extract shows only one band with the molecular weight of actin reacting with an actin antibody. (Adapted from Spudich *et al.*, 1988.)

unfertilized egg cortex was provided by Spudich *et al.* (1988) and Bonder *et al.* (1989). Spudich *et al.* (1988) examined the actin distribution in the unfertilized egg cortex by immunofluorescence microscopy, immunoelectron microscopy, and transmission electron microscopy, while Bonder *et al.* (1989) used cryosectioning techniques coupled with fluorescence and other forms of light microscopy. Electron microscopy of whole eggs and eggs fixed and permeabilized with Triton X-100 show filamentous actin in the prefertilization microvilli and in the cortex in close apposition to the plasma membrane, confirming earlier results discussed above. Antibodies specific to actin (Fig. 1) reveal a distinct 1-μm thick shell of actin in the unfertilized egg cortex (Fig. 2). This actin shell extends

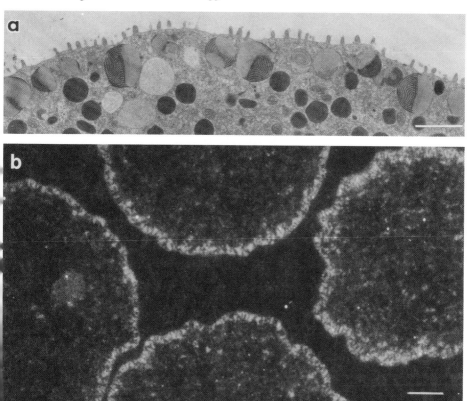

Fig. 2   Ultrastructure and actin distribution in the unfertilized egg cortex. (a) Thin-section electron micrograph showing the prefertilization microvilli at the egg surface and the cortical granules arranged in a monolayer close to the plasma membrane. Filamentous actin is not readily detected in the cell cortex. Bar = 2 μm. (b) Actin antibody staining reveals a 1-μm thick shell of concentrated actin in the cortex. This concentrated actin shell, which extends through the region containing the cortical granules, includes the filamentous and nonfilamentous actin. Eggs were fixed in 0.5% glutaraldehyde and 2% paraformaldehyde and embedded in Lowicryl K4M prior to sectioning for fluorescence microscopy. Bar = 2.5 μm. (Composite figure from Spudich *et al.*, 1988.)

beyond the filaments in the prefertilization microvilli and the region of the cortex immediately adjacent to the plasma membrane. The 1-μm thick actin shell includes the region containing the cortical granules. The actin surrounding the cortical granules does not bind phalloidin, confirming the electron microscopy data, which indicates that this actin is not filamentous.

Measurement of the elastic properties of the egg has suggested the existence of a gellike cortex (for review, see Hiramoto, 1970). Of particular interest in relation to the results reported above are the early physical measurements of Mitchison and Swann (1955) and Swann and Mitchison (1958), which led them to conclude that the sea urchin egg has a discrete cortical elastic shell of about 1.5 μm in thickness. The observation of a strikingly discrete cortical actin shell of similar dimensions fits well with their view of the egg cortex, and suggests that the cortical actin may contribute to the elasticity of the egg surface.

IV. Fertilization Induces Changes in the Organization of Actin in the Cortex

The egg cortex is a dynamic complex that participates actively in fertilization. Many excellent reviews summarize the extensive literature on the biochemical and ultrastructural changes that are initiated at fertilization in echinoderm eggs (Epel, 1978, 1982; Vacquier, 1981; Shapiro et al., 1981; Schatten, 1982). The major early cytoskeleton-associated events that occur on fertilization are the formation of the fertilization cone at the site of sperm binding, exocytosis of about 18,000 cortical granules, and the elongation of about 100,000 microvilli on the egg surface. All of these events are associated with obvious changes in the organization of the actin in the cortex.

A. Formation of the Fertilization Cone

The fertilization cone is an irregular extension of the egg cortex at the site of sperm–egg fusion. Within 1 min after insemination the fertilization cone can be seen to almost completely surround the fertilizing sperm, and by 2 min the fertilizing sperm has moved into the egg cortex. Parallel bundles of oriented actin filaments extend from the cone into the cytoplasm (Longo, 1980; Tilney and Jaffe, 1980). The polarity of the actin filaments in the cone are unidirectional; decoration of the filaments with muscle subfragment 1 (S1) gives rise to arrowheads pointing toward the cell center. Cytochalasin B inhibits formation of the fertilization cone and the uptake of the sperm, suggesting an active role for actin in the process (Longo, 1980; Schatten and Schatten, 1980).

B. Exocytosis of Cortical Granules and Assembly of Postfertilization Microvilli

Fertilization initiates the exocytosis of approximately 18,000 cortical granules (Schroeder, 1979), resulting in changes in the surface topography (Eddy and Shapiro, 1976) of the egg. Cortical granule exocytosis is initiated at about 25 sec after sperm binding and propagates from the site of sperm binding across the egg. The propagation results from a transient increase in internal free Ca^{2+} levels (for review, see Schuel, 1978; Vacquier, 1981). Within 45 sec after insemination exocytosis of the cortical granules is complete.

Elongation of microvilli and uptake of excess membrane by increased endocytic activity through coated pits at the base of the microvilli (Fischer and Rebhun, 1983; Sardet, 1984) also contribute to dramatic restructuring of the egg surface (Schroeder, 1979) after fertilization. A massive and highly synchronous actin polymerization occurs when microvilli elongate. The dynamics of the assembly of the microvillus have been examined by several investigators using a variety of techniques (Burgess and Schroeder, 1977; Tilney and Jaffe, 1980; Chandler and Heuser, 1981; Sardet, 1984). Burgess and Schroeder (1977) showed that each egg microvillus is made up of about seven actin filaments that are polarized with the arrowheads of myosin-decorated filaments pointing away from the microvillar tip. A cap of insoluble material is retained at the tip of each microvillus after extraction with Triton X-100, suggesting some specialized complex at the microvillus tip. Optical diffraction analysis of electron microscope images by Spudich and Amos (1979) revealed cross-bridges about every 130 Å that join neighboring actin filaments. The structure is similar to that of the hexagonally packed bundles of actin prepared *in vitro* by Kane (1975) from high-speed supernatants of sea urchin egg extracts and analyzed by DeRosier *et al.* (1977).

The data available show that the formation of actin-containing structures formed at fertilization, the fertilization cone or the microvilli that cover the egg surface, occurs as a two-step process, as suggested by Tilney and Jaffe (1980). First actin filaments are polymerized and are randomly oriented with one end attached to the plasma membrane. The filaments are then bound together, probably by association of specific bundling proteins. A composite image of the kinetics of microvilli assembly as derived from numerous studies is as follows. Following the completion of cortical granule exocytosis (1 min postfertilization) irregular protrusions containing actin filaments appear on the egg surface. By 2 min a large number of filaments, probably representing newly assembled actin, is present in the cortex. Very few filament bundles are present but many of the filaments contact the plasma membrane at one end. Between 3 and 5 min typical-looking microvilli are present at the cell surface. By 5 min the filaments become progressively more bundled and decoration by S1 shows that filaments are unidirectionally polarized with arrowheads pointing away from the plasma membrane. The possibility that the actin filaments are polymerized from specific areas on the plasma membrane was proposed by Tilney and Jaffe (1980). The short actin filament bundles in the prefertilization microvilli could be the nuclei for the polymerization of the actin filaments in the newly elongating microvilli.

Biochemical experiments by Dufresne *et al.* (1987) on the kinetics of actin assembly show fluctuations in the available G-actin pool in eggs immediately postfertilization. The G-actin pool in eggs was quantitated using the DNase-1 inhibition assay during the first 5 min after fertilization. The concentration of G-actin decreased between 0 and 30 sec after fertilization, possibly due to a small-scale assembly of actin. It is unclear whether this drop in the G-actin pool could be accounted for by the very rapid assembly of filaments in the fertilization cone. The G-actin content then increases between 30 and 60 sec, and this increase is dependent on a rise in intracellular Ca^{2+}. This increase may be generated from the concentrated form of nonfilamentous actin in the cortical shell of the unfertilized egg. The concentration of G-actin in eggs then decreases significantly below that of prefertilization levels, and this decrease probably accounts for the extensive assembly of F-actin seen by electron microscopy. Their data also suggest that the third phase of the postfertilization actin response, namely the extensive assembly, is dependent on cytoplasmic alkalinization. This result is consistent with the report of Begg and Rebhun (1979) that actin assembly can be induced in isolated unfertilized egg cortices by an increase in pH.

Thus, the two assembly states of actin in the unfertilized egg cortex may have distinct functions. The filamentous actin may provide the nuclei while the actin around the granules may be a storage form held in the cortex for the rapid assembly of the fertilization cone and the postfertilization microvilli. The concentrated pool of nonfilamentous actin in the cortex may provide the major source for the large increase in F-actin associated with these structures.

Another function of the cortical actin may be to regulate cortical granule exocytosis. Cortical actin has been proposed to be a barrier to exocytosis in neurosecretory cells (Linstedt and Kelly, 1987). Both populations of actin in the sea urchin cortex may be involved in sequestering the cortical granules and thereby serving as a barrier to vesicle fusion. Evidence from work of Lelkes *et al.* (1986) has suggested that depolymerization of the cortical actin filaments facilitates secretion in chromaffin cells. The findings that elevated Ca^{2+} levels in the lysis buffer can release actin associated with the isolated egg cortex (Spudich and Spudich, 1979) and that Ca^{2+} causes fusion of granules in an isolated cortical lawn (Vacquier, 1975) suggest a role for actin reorganization in granule fusion.

V. A Possible Role for Actin-Binding Proteins in the Organization of the Unfertilized Egg Actin

It is reasonable to speculate that a complex of actin and an actin-binding protein, which can be dissociated as a result of the increase in pH or Ca^{2+} that occurs at fertilization, is sequestered in the cortex. Proteins similar to those found in other eukaryotic cells that regulate the assembly state of actin have been purified and characterized from sea urchin eggs (for reviews, see Mabuchi and Hosoya, 1983; Mabuchi *et al.*, 1985; Mabuchi, 1986). However, their *in vivo* roles in the organization and restructuring of the actin in the egg cortex remain largely unknown. Of the many actin-binding proteins isolated from sea urchin eggs a 20-

kDa protein, actolinkin, purified from the insoluble fraction of unfertilized egg lysates, is one possible candidate for the sequestering of actin in the unfertilized egg cortex. It was isolated as a 1:1 complex with actin and caps the barbed end of actin filaments (Ishidate and Mabuchi, 1988). By immunoblot analysis it is found to be located in the isolated cortex fraction before fertilization. The distribution of the protein as detected by immunofluorescence in isolated fragments of the cortex did not change significantly after fertilization. Bonder *et al.* (1989) have shown the presence of spectrin in different domains of the unfertilized egg cortex. As was previously shown by Schatten *et al.* (1986), spectrin is found in association with the filaments present in the cortex. In addition, spectrin staining is found in the region of the nonfilamentous actin and around the cortical granules. The possibility that a spectrin-like protein could function in organizing and sequestering actin into a nonfilamentous state was proposed earlier by Tilney (1976).

Several Ca^{2+}-sensitive actin regulatory proteins have been isolated from sea urchin eggs and could play a role in the restructuring of actin that occurs at fertilization. A 45-kDa actin-binding protein that severs F-actin and caps the barbed ends of filaments in the presence of micromolar Ca^{2+} was purified from unfertilized egg extracts by Wang and Spudich (1984). Energy transfer and sedimentation experiments indicate that there is a net disassembly of actin filaments and an increase in the steady state nonfilamentous actin concentration in the presence of Ca^{2+} and the 45-kDa actin-binding protein. Biochemical characterization has shown that this protein is similar to severin, the Ca^{2+}-sensitive actin-severing protein form *Dictyostelium discoideum*. The 45-kDa actin-binding protein was isolated from egg extracts as a 1:1 complex with actin by Hosoya and Mabuchi (1984). A 100-kDa protein that fragments actin filaments in the presence of Ca^{2+} was purified from unfertilized eggs by Hosoya *et al.* (1986). This protein can form a complex with two or three actin monomers and may be a candidate for sequestering actin into a complex in the unfertilized egg cortex. Further biochemical and immunocytochemical localization work is needed to test this possibility. These proteins, when activated by elevated Ca^{2+} levels induced at fertilization, could release monomers from the cortical actin.

A wave of free Ca^{2+} precedes the exocytosis of cortical granules and the elongation of the actin-containing microvilli in sea urchin eggs (Epel, 1982). The fertilization-induced Ca^{2+} transient is initiated within seconds and levels are elevated from a resting Ca^{2+} concentration of about 150 nM to 2 μM after fertilization (Poenie *et al.*, 1985). Calcium ion causes cortical granule fusion on isolated fragments (Vacquier, 1975) of the egg cortex (Steinhardt *et al.*, 1977) and is likely to be one trigger for the fertilization-induced reorganization of actin in the cortex.

Increase in the concentration of free Ca^{2+} in lysis buffer results in solubilization of actin from isolated cortices of unfertilized sea urchin eggs (Fig. 3). When fragments of the unfertilized egg cortex are prepared in the presence of 2 mM Ca^{2+} and labeled with actin antibody, the actin remaining in the cortex is

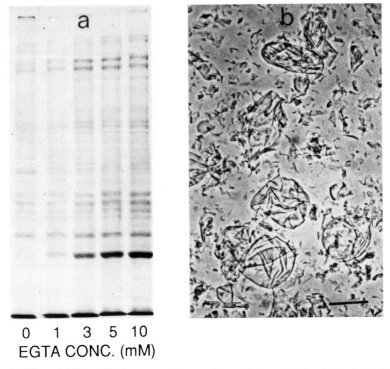

0   1   3   5   10
EGTA CONC. (mM)

Fig. 3 Effect of different EGTA concentrations on the associations of actin with the isolated egg cortex. (a) SDS gel patterns of Triton-extracted unfertilized egg cortices isolated in different concentrations of EGTA. The major Coomassie-staining band is actin. (b) Phase micrograph of Triton-extracted unfertilized cortex fragments isolated in 5 mM EGTA. Bar = 100 μm. (Adapted from Spudich and Spudich, 1979.)

distributed in weakly fluorescent punctate spots (Spudich, 1988). These spots may represent the assembled actin detected by electron microscopy in the prefertilization microvilli. A population of actin filaments that is not depolymerized in response to the Ca^{2+} wave at fertilization could provide the nuclei for actin polymerization, as was suggested by Tilney (1976) for the echinoderm sperm. Actin released from the cortex in response to the Ca^{2+} wave could polymerize onto these nuclei for the rapid assembly of the fertilization cone or microvilli.

VI. Summary and Future Perspectives

The molecular basis of the cortical changes in actin that occur on fertilization remains to be elucidated. The discovery of two distinct assembly states of actin in the prefertilization cortex and the purification of numerous actin-associated proteins that are likely to regulate these cortical changes provide appropriate materials for future studies. One problem has been the difficulty in properly fixing the rather large sea urchin eggs for localization studies. The applications of cryofixation techniques and Lowicryl embedding methods, however, now allow the localization of the various actin-associated proteins as a function of time after fertilization. Such structural studies should provide important clues regarding the possible involvement of these proteins in the fertilization-induced changes in actin organization in the sea urchin egg cortex.

Acknowledgment

The author gratefully acknowledges James A. Spudich for critical reading of this manuscript and for challenging discussions.

References

Begg, D. A., and Rebhun, L. I. (1979). pH regulates the polymerization of actin in the sea urchin egg cortex. *J. Cell Biol.* **83**, 241–248.

Bonder, E. M., Fishkind, D. J., Cotran, N. M., and Begg, D. A. (1989). The cortical actin-membrane cytoskeleton of unfertilized sea urchin eggs: Analysis of the spatial organization and relationship of filamentous actin, nonfilamentous actin, and egg spectrin. *Dev. Biol.* **134**, 327–341.

Burgess, D. R., and Schroeder, T. E. (1977). Polarized bundles of actin filaments within microvilli of fertilized sea urchin eggs. *J. Cell Biol.* **74**, 1032–1037.

Chandler, D. E., and Heuser, J. (1981). Postfertilization growth of microvilli in the sea urchin egg: New views from eggs that have been quickfrozen, freeze-fractured, and deeply etched. *Dev. Biol.* **82**, 393–400.

Cline, C. A., and Schatten, G. (1986). Microfilaments during sea urchin fertilization: Fluorescence detection with rhodaminyl phalloidin. *Gamete Res.* **14**, 277–291.

Conklin, E. G. (1917). Effects of centrifugal forces on the structure and development of the eggs of *Crepidula*. *J. Exp. Zool.* **22**, 311–419.

Davidson, E. H. (1986). "Gene Activity in Early Development." Academic Press, Orlando, Florida.

DeRosier, D. J., Mandelkow, E., Silliman, A., Tilney, L., and Kane, R. (1977). Structure of actin containing filaments from two types of nonmuscle cells. *J. Mol. Biol.* **113**, 679–695.

Dufresne, L., Swezey, R. R., and Epel, D. (1987). Kinetics of actin assembly attending fertilization or artificial activation of sea urchin eggs. *Exp. Cell Res.* **172**, 32–42.

Eddy, E. M., and Shapiro, B. M. (1976). Changes in the topography of sea urchin egg after fertilization. *J. Cell Biol.* **71**, 35–48.

Epel, D. (1978). Mechanisms of activation of sperm and egg during fertilization of sea urchin gametes. *Curr. Top. Dev. Biol.* **12**, 185–246.

Epel, D. (1982). The physiology and chemistry of calcium during the fertilization of eggs. *In* "Calcium and Cell Function" (W. Y. Cheung, ed.), Vol. 11, pp. 356–383. Academic Press, New York.

Fischer, G. W., and Rebhun, L. I. (1983). Sea urchin egg cortical granule exocytosis is followed by a burst of membrane retrieval via uptake into coated vesicles. *Dev. Biol.* **99**, 456–472.

Henson, J. H., and Begg, D. A. (1988). Filamentous actin organization in the unfertilized sea urchin egg cortex. *Dev. Biol.* **127,** 338–348.

Hiramoto, Y. (1970). Rheological properties of sea urchin eggs. *Biorheology* **6,** 201–234.

Hosoya, H., and Mabuchi, I. (1984). A 45,000 MW protein–actin complex from unfertilized sea urchin eggs affects the assembly properties of actin. *J. Cell Biol.* **99,** 994–1001.

Hosoya, H., Mabuchi, I., and Sakai, H. (1986). A 100-kDa Ca^{2+}-sensitive actin-fragmenting protein from unfertilized sea urchin egg. *Eur. J. Biochem.* **154,** 233–239.

Ishidate, S., and Mabuchi, I. (1988). Localization and possible function of 20 kDa actin-modulating protein (actolinkin) in the sea urchin egg. *Eur. J. Cell Biol.* **46,** 275–281.

Kane, R. E. (1975). Preparation and purification of polymerized actin from sea urchin egg extracts. *J. Cell Biol.* **66,** 305–315.

Lelkes, P. I., Friedman, J. E., Rosenheck, K., and Oplatka, A. (1986). Destabilization of actin filaments as a requirement for the secretion of catecholamines from permeabilized chromaffin cells. *FEBS Lett.* **208,** 357–363.

Linstedt, A. D., and Kelly, R. B. (1987). Overcoming barrier to exocytosis. *Trends Neurosci.* **10,** 446–448.

Longo, F. J. (1980). Organization of microfilaments in sea urchin (*Arbacia punctulata*) eggs at fertilization: Effects of cytochalasin B. *Dev. Biol.* **74,** 422–433.

Mabuchi, I. (1986). Biochemical aspects of cytokinesis. *Int. Rev. Cytol.* **101,** 175–213.

Mabuchi, I., and Hosoya, H. (1983). Actin-modulating proteins in echinoderm eggs. *In* "Actin: Structure and Function in Muscle and Nonmuscle Cells" (C. G. Dos Remedios and J. A. Braden, eds.), pp. 309–316. Academic Press, Sydney, Australia.

Mabuchi, I., and Sakai, H. (1972). Cortex protein of sea urchin eggs, its purification and other protein components of the cortex. *Dev., Growth Differ.* **14,** 247–261.

Mabuchi, I., and Spudich, J. A. (1980). Purification and properties of soluble actin from sea urchin eggs. *J. Biochem. (Tokyo)* **87,** 785–802.

Mabuchi, I., Hosoya, H., Ishidate, S., Hamaguchi, Y., Tsukita, S., and Tsukita, S. (1985). Actin-modulating proteins in echinoderm eggs. II. *In* "Cell Motility: Mechanism and Regulation" (H. Ishikawa, S. Hatano, and H. Sato, eds.), pp. 175–188. Univ. of Tokyo Press, Tokyo.

Mitchison, J. M., and Swann, M. M. (1955). The mechanical properties of the cell surface: The sea urchin egg from fertilization to cleavage. *J. Exp. Biol.* **32,** 734–750.

Poenie, M., Alderton, J., Tsien, R. Y., and Steinhardt, R. A. (1985). Changes of free calcium levels with stages of the cell division cycle. *Nature (London)* **315,** 147–149.

Sardet, C. (1984). The ultrastructure of the sea urchin egg cortex isolated before and after fertilization. *Dev. Biol.* **105,** 196–210.

Schatten, G. (1982). Motility during fertilization. *Int. Rev. Cytol.* **79,** 59–92.

Schatten, H., and Schatten, G. (1980). Surface activity at the egg plasma membrane during sperm incorporation and its cytochalasin B sensitivity. *Dev. Biol.* **78,** 435–449.

Schatten, H., Cheney, R., Balczon, R., Willard, M., Cline, C., Simerly, C., and Schatten, G. (1986). Localization of fodrin during fertilization and early development of sea urchin and mice. *Dev. Biol.* **118,** 457–477.

Schroeder, T. (1979). Surface area change at fertilization: Resorption of the mosaic membrane. *Dev. Biol.* **73,** 306–327.

Schuel, H. (1978). Secretory functions of egg cortical granules in fertilization and development: A critical review. *Gamete Res.* **1,** 299–382.

Shapiro, B. M., Schackmann, R. W., and Gabel, C. A. (1981). Molecular approaches to the study of fertilization. *Annu. Rev. Biochem.* **50,** 815–843.

Spudich, A. (1988). Organization of actin in the cell cortex. Ph.D. Thesis, Stanford University, Stanford, California.

Spudich, A., and Spudich, J. A. (1979). Actin in Triton-treated cortical preparations of unfertilized and fertilized sea urchin eggs. *J. Cell Biol.* **82,** 212–226.

Spudich, A., Giffard, R. G., and Spudich, J. A. (1982). Molecular aspects of cortical actin filament formation upon fertilization. *Cell Differ.* **11**, 281–284.

Spudich, A., Wrenn, J. T., and Wessells, N. K. (1988). Unfertilized sea urchin eggs contain a discrete cortical shell of actin that is subdivided into two organizational states. *Cell Motil. Cytoskel.* **9**, 85–96.

Spudich, J. A., and Amos, L. A. (1979). Structure of actin filament bundles from microvilli of sea urchin eggs. *J. Mol. Biol.* **129**, 319–331.

Steinhardt, R. A., Zucker, R., and Schatten, G. (1977). Intracellular calcium release at fertilization in the sea urchin egg. *Dev. Biol.* **58**, 185–196.

Swann, M. M., and Mitchison, J. M. (1958). The mechanism of cleavage in animal cells. *Biol. Rev. Cambridge Philos. Soc.* **88**, 103–135.

Tilney, L. G. (1976). The polymerization of actin. III. Aggregates of nonfilamentous actin and its associated proteins: A storage form of actin. *J. Cell Biol.* **69**, 73–89.

Tilney, L. G. (1978). Polymerization of actin. V. A new organelle, the actomere, that initiates the assembly of actin filaments in *Thyone* sperm. *J. Cell Biol.* **77**, 551–564.

Tilney, L. G., and Jaffe, L. A. (1980). Actin, microvilli and the fertilization cone of sea urchin eggs. *J. Cell Biol.* **87**, 771–782.

Vacquier, V. D. (1975). The isolation of intact cortical granules from sea urchin eggs: Calcium ions trigger granule discharge. *Dev. Biol.* **43**, 62–74.

Vacquier, V. D. (1981). Dynamic changes of the egg cortex. *Dev. Biol.* **84**, 1–26.

Vacquier, V. D., and Moy, G. W. (1980). The cytolytic isolation of the cortex of the sea urchin egg. *Dev. Biol.* **77**, 178–190.

Wang, L.-L., and Spudich, J. A. (1984). A 45,000-mol-wt protein from unfertilized sea urchin eggs severs actin filaments in a calcium-dependent manner and increases the steady-state concentration of nonfilamentous actin. *J. Cell Biol.* **99**, 844–851.

Yonemura, S., and Mabuchi, I. (1987). Wave of cortical actin polymerization in the sea urchin egg. *Cell Motil. Cytoskel.* **7**, 46–53.

2

Localization of *bicoid* Message during *Drosophila* Oogenesis

Edwin C. Stephenson* and Nancy J. Pokrywka
Department of Biology
University of Rochester
Rochester, New York 14627

I. Introduction

In the past few years it has become clear that the determinant nature of *Drosophila* development results from molecular asymmetries present in the egg before fertilization. There are only a few such positional markers in the egg, specifying anterior and posterior poles and the dorsal–ventral axis (Anderson, 1987; Nüsslein-Volhard *et al.*, 1987). During early embryogenesis a complex cascade of genetic regulatory interactions refines this coarse positional information into the detailed pattern of the embryo (Akam, 1987). The molecular asymmetry that marks the anterior end of *Drosophila* eggs is the *bicoid* messenger RNA (Berleth *et al.*, 1988). *bicoid* message is produced only during oogenesis, and is stably localized at one end of the oocyte, egg, and early embryo. The signal for localization within *bicoid* message appears to be the secondary structure of the 3′ untranslated end (Macdonald and Struhl, 1988; Macdonald, 1990). *bicoid* protein is translated during early embryogenesis, is restricted to the anterior embryo, and activates the transcription of one or more anterior-specific genes in the zygote (Driever and Nüsslein-Volhard, 1988, 1989; Struhl *et al.*, 1989).

Of the many interesting features of this system, we restrict the following discussion to the question of the cellular mechanisms that lead to the localization of *bicoid* message in the anterior oocyte and egg. For instance, what components of the cytoskeleton are involved and what do they do? Why is *bicoid* message localized specifically in the cortical cytoplasm of the anterior egg: is cytoplasm in the cortex or in the anterior egg different at the molecular level from cytoplasm elsewhere? How does *bicoid* message move from its site of synthesis to its

---

* Present address: Department of Biology, University of Alabama, Tuscaloosa, Alabama, 35487.

Current Topics in Developmental Biology, Vol. 26
Copyright © 1992 by Academic Press, Inc. All rights of reproduction in any form reserved.

localized position: is movement active or passive, and are movement and localization linked in any way? Does *bicoid* message localization require a large set of unusual molecules and organelles, or is localization simply a specialization of a common cellular process?

Oogenesis in *Drosophila*, as in most insects, is meroistic (for review, see Mahowald and Kambysellis, 1980). This means, in effect, that none of the molecules ending up in the egg is a product of the oocyte genome itself. The primary protein component of any egg, yolk, is made by somatic cells and taken into the oocyte by endocytosis. All other protein and RNA components of the egg are synthesized in nurse cells. Each egg chamber contains 16 cells of germline origin, generated by four mitotic divisions early in oogenesis. One of the 16 cells is the oocyte, and the other 15 are nurse cells. Cytokinesis during the four mitoses that generate the oocyte–nurse cell complex is incomplete, the result being that nurse cells and the oocyte remain connected by intercellular bridges (also called ring canals). The oocyte nucleus appears to be transcriptionally inactive; nurse cell nuclei, on the other hand, become highly polyploid and are active in transcription. The products of nurse cells pass into the oocyte in two distinct phases. In previtellogenic and vitellogenic stages of oogenesis, movement from nurse cells into the oocyte is slow and involves mechanisms that are poorly understood and in dispute (Bohrmann and Gutzeit, 1987; Woodruff *et al.*, 1988). In postvitellogenic stages, nurse cell cytoplasm floods into the oocyte in a bulk flow process, and the residue of nurse cells disintegrates. Each egg chamber contains a third cell type, follicle cells, that are not in cytoplasmic continuity with the nurse cell–oocyte complex, and do not make *bicoid* message or appear to play a role in its localization.

Oogenesis has been divided into 14 stages (see King, 1970); however, the rough stages described in the preceding paragraph are sufficient for the discussion that follows: in previtellogenic stages nurse cells and the oocyte grow at approximately equal rates, in vitellogenic stages the oocyte incorporates yolk and thus becomes larger and clearly distinct from its nurse cell sisters, and in postvitellogenic stages nurse cells dump their cytoplasm into the oocyte and degenerate.

II. Cell Biology of *bicoid* Message Localization

Most available information on the mechanism of *bicoid* message localization is inferred from *in situ* hybridization experiments like those shown in Figs. 1 and 2 (Berleth *et al.*, 1988; Stephenson *et al.*, 1988; St. Johnston *et al.*, 1989). *bicoid* message is detectable at all stages of oogenesis except the very earliest, and is present in both nurse cells and the oocyte. In early stages of oogenesis the distribution of *bicoid* message in nurse cells is patchy, with some regions of nurse cell cytoplasm containing higher levels than other regions (Fig. 1a and b). In

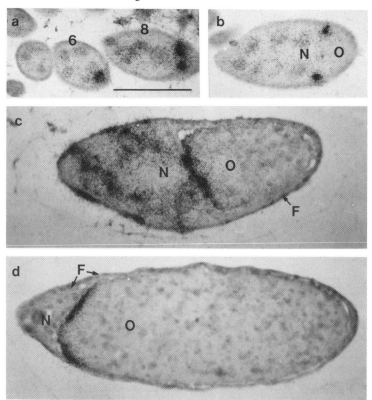

Fig. 1 *bicoid* message localization during oogenesis. *Drosophila* egg chambers were fixed, sectioned, and hybridized with a radioactive probe to detect *bicoid* message. The nurse cell complex and the anterior oocyte are to the left in each panel. The bar in (a) is 100 μm for the entire figure. Numbers are the stages of King (1970). N, Nurse cell complex; O, oocyte; F, follicle cells. (a) Previtellogenic (stage 6) and early vitellogenic (stage 8) egg chambers. Note the uneven distribution of *bicoid* message in the nurse cell complex and the accumulation of *bicoid* message in the oocyte at the posterior tip of the egg chamber. (b) Vitellogenic (stage 9) egg chamber. *bicoid* message forms a peripheral ring in the oocyte (see Fig. 2c); a section through the ring appears as two peripheral patches. (c) Late vitellogenic egg chamber (stage 10). Note the strong localization of *bicoid* message around nurse cell nuclei and the broadening of *bicoid* message to cover the entire anterior oocyte cortex. (d) Nearly mature oocyte (stage 12). Nurse cell remnants and anterior follicle cells form an anterior cap at the tip of the oocyte.

vitellogenic stages *bicoid* message accumulates strongly around nurse cell nuclei (Fig. 1c). *bicoid* message is, of course, also localized in the oocyte. In early vitellogenic stages, *bicoid* message accumulates in several discrete patches in peripheral regions of the anterior margin (Fig. 2a), and slightly later these patches have merged into a peripheral ring (Fig. 2c). In vitellogenic stages *bicoid* message is concentrated in a more uniform mass, in a central part of the anterior

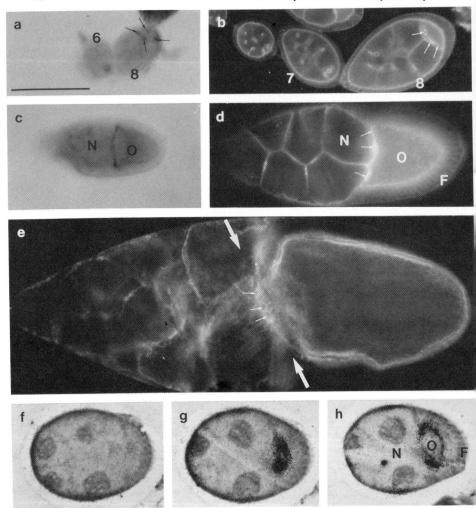

Fig. 2 *bicoid* message is localized in cortical oocyte cytoplasm. (Abbreviations are as in Fig. 1.) Bar in (a) is 100 μm for the entire figure. All figures here are printed at the same magnification; however, dehydrated tissues (a, c, f, g, and h, and all shown in Fig. 1) are considerably shunken from their live dimensions. (a and c) Whole egg chambers were hybridized with digoxygenin-labeled *bicoid* probe, and hybrids detected by horseradish peroxidase staining. (b, d, and e) Whole egg chambers stained with rhodamine-labeled phalloidin to detect F-actin. Intercellular bridges appear as bright rings. Note the cortical distribution of actin in nurse cells and the oocyte, and larger amounts of actin in the oocyte. In the late vitellogenic egg chamber in (e), nurse cell actin filaments are reorganized in preparation for squeezing nurse cell cytoplasm into the oocyte. (f, g, and h) Three adjacent grazing sections through a vitellogenic egg chamber, hybridized with radioactively labeled *bicoid* probe. In early vitellogenic egg chambers (stage 8), *bicoid* message accumulates in four discrete patches (arrows in a) in approximately the same positions as the four bridges between nurse cells and the oocyte (arrows in b). In early vitellogenic egg chambers (stage 9), *bicoid* message

2. Localization of *bicoid* Message 27

oocyte margin (Fig. 1c). As nurse cells degenerate and dump their cytoplasmic contents into the oocyte, *bicoid* message remains localized at the anterior margin of the oocyte (Fig. 1d). Nurse cells of late vitellogenic egg chambers contain large amounts of *bicoid* message that is dumped into the oocyte with the rest of the nurse cell cytoplasm in late oogenesis, and appears to be localized during this process.

What can we learn of the dynamic process of *bicoid* message localization from the admittedly static picture offered by *in situ* hybridization? Like other RNAs, *bicoid* message is transcribed predominantly or exclusively in nurse cell nuclei. While a small amount of transcription from the oocyte nucleus cannot be excluded, most *bicoid* message is clearly synthesized in nurse cells. The asymmetric accumulation around nurse cell nuclei is intriguing. What is *bicoid* message localized in this way, and to what underlying cellular structure might it be attached? *bicoid* message is unusual in being localized in the oocyte, so a reasonable guess might be that nurse cell localization and oocyte localization are part of the same process, and that a component of the cellular machinery is asymmetrically disposed in nurse cell cytoplasm. St. Johnston *et al.* (1989) interpreted the position of *bicoid* message as the apical side of the nurse cell, and suggested that the apical–basal polarity of nurse cells corresponds to the anterior–posterior polarity of the oocyte. Is there an obvious structure or cellular organelle to which *bicoid* message might be attached? Microtubules in nurse cells have been shown to be primarily perinuclear (Gutzeit, 1986a), raising the interesting possibility that *bicoid* message is attached to this component of the cytoskeleton immediately after synthesis. It is not certain that microtubules are asymmetrically arranged around nurse cell nuclei in the pattern of *bicoid* message; however, *bicoid* message might be associated with an asymmetric subset. The Golgi complex in most cells shows a similar asymmetric perinuclear distribution, but nothing of what we know or have reason to expect of the Golgi complex suggests its participation in RNA localization. Masses of fibrous body material, composed of protein and RNA, are associated with the cytoplasmic face of nurse cell nuclei (Mahowald and Strassheim, 1970), and pieces of this material eventually migrate into the oocyte. Is it possible that *bicoid* message is incorporated into the fibrous body material just outside nurse cell nuclei, and localized in this form in the anterior oocyte? Mahowald (1971) suggested that the fibrous body material is used to

---

accumulates as a peripheral ring (c) at a stage when the four bridges between nurse cells and the oocyte are still peripheral (arrows in d). The egg chamber in (d) is slightly older than that in (c), but much of the difference in size is due to the shrinkage of (c). In late vitellogenic egg chambers bridges connect to the central anterior margin of the oocyte (e). At this stage *bicoid* message accumulates in this central location. Large white arrows in (e) indicate the approximate orientation of the sections in panels (f), (g), and (h). The section in (f) passes through the extreme margin of nurse cells and does not include large amounts of *bicoid* message. The next 5-μm section (g) grazes the margin of the oocyte, including the mass of *bicoid* message, and the third section (h) is slightly deeper within the oocyte and beyond the major part of the mass of *bicoid* message.

store maternal RNAs for embryogenesis, and part of the fibrous body material is localized to the posterior pole of the egg, comprising the polar granules that appear to induce pole cell development (Mahowald, 1971; Hay *et al.*, 1988). Not all of the fibrous body material is localized to the posterior pole, suggesting that some of it has a function other than polar granule formation.

Stage-specific changes in the position of *bicoid* message accumulation in the oocyte were pointed out by St. Johnston *et al.* (1989), who noted that *bicoid* message accumulates in a peripheral ring in early vitellogenic (stage 8) oocytes. We have carried out similar experiments, and observe that at a slightly earlier stage in the process *bicoid* message accumulates as four discrete peripheral patches (Fig. 2a). Four of the 15 nurse cells have direct connections to the oocyte, and the position of the 4 intercellular bridges correlates well with the position of *bicoid* message in the anterior oocyte. In early stages, when *bicoid* message is localized in the oocyte as four peripheral patches or a peripheral ring, the bridges from the four proximal nurse cells are connected to the peripheral margins of the anterior oocyte (Fig. 2b and d; also see Warn *et al.*, 1985). As the oocyte grows, these four bridges remain close together, so that in vitellogenic stage egg chambers these four bridges connect nurse cells to a more central region of the anterior oocyte margin (Fig. 2e). At these stages *bicoid* message accumulates at a similar central anterior location (Fig. 1c). These observations suggest that the attachment points for *bicoid* message lie just within the oocyte, in the cortical cytoplasm just across the intercellular bridges connecting the four proximal nurse cells to the oocyte. The restriction of *bicoid* message to cortical oocyte cytoplasm is also obvious in *in situ* hybridizations to adjacent sections that graze the mass of localized *bicoid* message (Fig. 2f–h), and suggest that *bicoid* message is restricted to a cytoplasmic layer that is less than 5 μm thick, the thickness of sections used in these experiments.

What are the points of attachment for *bicoid* message in cortical oocyte cytoplasm? We accept as an article of faith the idea that some component of the cytoskeleton must anchor *bicoid* message in the anterior oocyte, but available evidence is limited and not useful in deciding which component(s) might be important. Microtubules are abundant in both nurse cells and the oocyte but are not obviously enriched in the oocyte cortex (Gutzeit, 1986a). As mentioned above, microtubules are concentrated around nurse cell nuclei, and so could be involved in nurse cell localization. In vitellogenic egg chambers, actin filaments are abundant in the periphery of nurse cells and in the oocyte cortex (Fig. 2b, d, and e; Warn *et al.*, 1985; Gutzeit, 1986b); the latter is, of course, an appropriate location for a *bicoid* message-anchoring system. *Drosophila* egg chambers presumably contain intermediate filaments and a membrane cytoskeleton, but neither has been described.

In trying to envision the process of *bicoid* message localization, it is attractive to imagine it as a receptor–ligand interaction: on entering the oocyte, *bicoid* messenger RNA molecules are captured by receptors anchored in the anterior

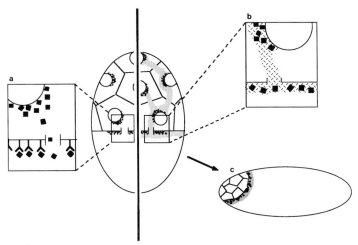

Fig. 3 Two models for *bicoid* message localization. A schematic egg chamber is drawn in the center. Enlargement (a) shows a simple trapping model, with *bicoid* message (squares) captured by receptors anchored in the cortical cytoplasm. In (b), *bicoid* message is attached to a hypothetical network (stippled) running through the oocyte–nurse cell complex. The entire network, with *bicoid* message still attached, passes into the oocyte during the bulk flow phase, but remains localized at the anterior margin of late oocytes and eggs (c).

cortical cytoplasm (Fig. 3a). This interpretation accounts nicely for the observation that *bicoid* message accumulates in the thin layer of oocyte cytoplasm just across the intercellular bridges. The receptors themselves would not need to be restricted to the anterior oocyte, so long as they were in sufficient excess to ensure that all *bicoid* message is captured soon after it enters the oocyte. The position of localized *bicoid* message might result simply from the proximity of nurse cells, and not from any inherent structural difference in anterior cytoplasm. This scenario assumes that *bicoid* message is free to diffuse within the nurse cell–oocyte syncytium, eventually finding its anchored receptor in the anterior oocyte. However, information presented above suggests that *bicoid* message is not free to diffuse in nurse cells, because it is localized around the nurse cell nucleus just after synthesis. In addition, the localization of *bicoid* message during the bulk flow of nurse cell cytoplasm into the oocyte is difficult to explain as a simple receptor–ligand interaction.

Late in oogenesis the egg chamber undergoes a massive rearrangement. The cytoplasmic contents of nurse cells, approximately equal in volume to that of the oocyte at the beginning of the process, flow into the oocyte in the course of about an hour. The bulk flow process appears to be driven by the contraction of actin filaments, as nurse cells essentially squeeze their contents into the oocyte. The cytoplasmic stream moves at a rate of approximately 2 μm/sec, through bridges about 12 μm in diameter (Gutzeit and Koppa, 1982). *bicoid* message already present in the oocyte maintains its localized position throughout this chaotic process, suggesting its cytoplasmic attachment points are very stably anchored. More remarkably, *bicoid* message appears to be localized *during* the bulk flow process itself. This conclusion is slightly uncertain, because *in situ* hybridization is by its nature poorly suited for analyzing a dynamic process. Nonetheless, nurse cells in late vitellogenic egg chambers contain large quantities of *bicoid* message that presumably enter the oocyte with the rest of nurse cell cytoplasm. Because the total amount of *bicoid* message localized in the oocyte seems to increase in late oogenesis, the simplest conclusion is that it is localized during the bulk flow stage. We believe the localization of *bicoid* message during bulk flow rules out the simplest form of the receptor–ligand model outlined above. It is difficult to imagine that an anchored receptor could efficiently and specifically capture *bicoid* messenger ribonucleoproteins (RNPs) from the complex and rapidly moving stream of nurse cell cytoplasm flowing into the oocyte.

We envision two alternatives to the simple receptor–ligand model, seeking in both to account for the localization of *bicoid* message around nurse cell nuclei and for the localization of *bicoid* message in the oocyte during bulk flow. The first possibility retains most of the receptor–ligand idea, but supposes that *bicoid* message is assembled into or attached to a large macromolecular complex or cellular organelle in nurse cell cytoplasm. Candidates for macromolecular assemblies to which *bicoid* message might be attached include microtubules or other cytoskeletal components, or the fibrous body material that gathers around nurse cell nuclei. During bulk flow, *bicoid* message is swept into the oocyte attached to these large structures, in which form it might be more easily captured by anchored receptors in the anterior cortical cytoplasm.

The second possibility rejects the idea of a simple receptor in the oocyte cortex, and supposes that the localization machinery comprises an extensive cytoskeletal network that forms a continuous structure in nurse cells and is anchored in the anterior cytoplasm of the oocyte (Fig. 3b). The perinuclear concentration of *bicoid* message in nurse cells reflects its attachment to this hypothetical network. During bulk flow, this network is swept into the oocyte along with everything else, but unlike most nurse cell cytoplasm retains some of its organization and its anchored position in the anterior oocyte. The attachment of *bicoid* message to the network in nurse cell cytoplasm thus ensures its later localization in the oocyte. There is no evidence for the existence of an extensive nurse cell–oocyte network at the stages of oogenesis of interest here, but a structure with some of the expected characteristics exists in younger egg chambers. The polyfusome (Telfer, 1975; King *et al.*, 1982) is a fine fibrillar ground substance that replaces spindle microtubules following the four divisions that generate the oocyte and nurse cells, and forms an unbroken network through the 16 cells of the oocyte–nurse cell syncytium. The fusome ground substance has a transient existence, and can no longer be detected at stages of oogenesis relevant to *bicoid* message localization. It is not clear whether the level of supracellular organization represented by the polyfusome simply disappears in later stages or persists in a less recognizable form. The methods employed to characterize the polyfusome in early stages (differential hematoxylin staining and fine-structure studies by electron microscopy) might not reveal a similar structure or its successor in later stages, when the egg chamber is larger and more complex.

These models are not mutually exclusive, and it is conceivable that the mechanism of *bicoid* message localization is different at different stages of oogenesis. For instance, a simple receptor–ligand interaction might be sufficient early in oogenesis when entry into the oocyte is slow, with a more complex mechanism required during the bulk flow of nurse cell cytoplasm into the oocyte late in oogenesis. The strongly perinuclear accumulation of *bicoid* message is apparent only in vitellogenic stages of oogenesis, suggesting that this aspect of the localization machinery is stage specific. Perhaps perinuclear accumulation is an early step in the specialized process that localizes *bicoid* message during bulk flow, but is not part of localization in early stages of oogenesis.

III. Genes Required for *bicoid* Message Localization

We know of two genes whose products are involved in *bicoid* message localization in oogenesis, *exuperantia* and *swallow* (Frohnhöfer and Nüsslein-Volhard, 1987). Each is a maternal effect gene—embryonic development is defective because *bicoid* message is uniformly distributed instead of being localized at the anterior tip of eggs produced by homozygous females. The ultimate effect of mutations in *exuperantia* and *swallow* is the same, uniform distribution of *bicoid* message; however, the two genes are distinguishable in their effects on intermediate stages of *bicoid* message localization (Berleth *et al.*, 1988; St. Johnston *et al.*, 1989; N. J. Pokrywka and E. C. Stephenson, unpublished). In *exuperantia*$^-$ egg chambers *bicoid* message does not accumulate around nurse cell nuclei and is not localized in the oocyte at any stage of oogenesis. St. Johnston *et al.* (1989) reported an abnormal diffuse ring of *bicoid* message accumulation in previtellogenic oocytes, but we have not seen any evidence of localization. If we accept the idea that *bicoid* message attachment to a perinuclear organelle in nurse cell cytoplasm is an early step in localization, then the *exu*$^+$ gene product must be responsible for this attachment, and thus probably plays an early role in localization. In contrast, early aspects of *bicoid* message localization in *swallow*$^-$ egg chambers appear normal, suggesting that *swallow* plays a later role than *exuperantia* (Berleth *et al.*, 1988; Stephenson *et al.*, 1988; St. Johnston *et al.*, 1989). In previtellogenic and early vitellogenic *swallow*$^-$ egg chambers, *bicoid* message shows normal perinuclear accumulation in nurse cells and is correctly localized in the oocyte. However, localization deteriorates in a stepwise fashion in late vitellogenic stages of *swallow* oocytes. The strict anterior localization is lost first, with some *bicoid* message appearing in the lateral oocyte margin. The pattern of this deterioration continues, with *bicoid* message remaining mostly cortical but losing its anterior localization, and eventually losing even the cortical localization. These observations have been interpreted to suggest that *swallow* plays a role in maintaining the position or stability of the cytoskeletal elements to which *bicoid* message is presumably attached in the oocyte (Berleth *et al.*, 1988; Stephenson *et al.*, 1988). A role of this sort should be required relatively late in the process, and might account for the stepwise disintegration of the localized pattern.

In addition to its role in localizing *bicoid* message during oogenesis, *swallow* appears to play a role in nuclear divisions early in embryogenesis. In normal development, embryonic nuclei undergo a series of synchronous cleavages, eventually migrating to the periphery of the egg by the eighth cleavage. Five more synchronous cleavages occur at the periphery before cellular membranes partition the peripheral cytoplasm into cells. These early cleavage divisions are defective in several characteristic ways in *swallow*⁻ embryos (Zalokar *et al.*, 1975): (1) many embryos arrest after only a few cleavage divisions; (2) nuclear migration to the embryonic periphery is precocious, with some nuclei reaching the periphery as early as the fifth cleavage; (3) nuclear migration to the periphery is uneven, with some parts of the blastoderm unpopulated by cleavage nuclei; (4) nuclei often divide asynchronously; (5) the number of abnormal cleavages is high. These cleavage stage defects cannot be rationalized as resulting from mislocalized *bicoid* message, because *exuperantia* embryos do not show similar defects. Instead, we believe that the *swallow* gene product has a role in regulating embryonic cleavage divisions, in addition to its role in localizing *bicoid* message in oogenesis. We do not now know what role *swallow* plays in early development; however, the nature of the defects is most easily imagined as a role in organizing or regulating components of the cytoskeleton that are responsible for mitotic divisions and nuclear migration. Thus both aspects of the *swallow* phenotype, in oogenesis and early embryogenesis, are consistent with a cytoskeletal role.

The *swallow* gene has been cloned and sequenced (Stephenson *et al.*, 1988). *swallow* expression appears to be limited to the female germline, and like most mRNAs is present in both nurse cells and the oocyte. The mRNA persists for the first few hours of embryogenesis, then disappears and cannot be detected at any subsequent stage of development. The *swallow* messenger RNA is not spatially localized at any stage of oogenesis or early embryogenesis. Its developmentally limited expression might mean that *swallow* carries out a role that is required only in oogenesis or early embryogenesis. As an alternative, the *swallow* protein might carry out a function common to many stages, with *swallow* required in oogenesis and early embryogenesis, and the product of another gene carrying out a similar function in other stages of development. We have looked without success for *swallow* homologs in the *Drosophila* genome by low-stringency hybridization, suggesting this hypothetical homolog does not exist or is too diverged in nucleotide sequence to be recognizable by this method.

The protein predicted from the gene sequence has not been helpful in deciding what role *swallow* might play in oogenesis and early embryogenesis (Chao *et al.*, 1991). Secondary structure predictions indicate the presence of an amphipathic helix in the middle of the protein. This region is similar in structure to domains that have been shown to mediate protein–protein interactions in other systems, particularly transcription factor proteins (Kouzarides and Ziff, 1989). We do not now know the identity of any proteins with which *swallow* might be associated. The predicted *swallow* protein also contains a domain with similarity to the consensus RNA-binding domain (Bandziulis *et al.*, 1989; Query *et al.*, 1989). The similarity is low enough that we are not certain of its significance. As described above, our best guess for the function of *swallow* is that it regulates or stabilizes cytoskeletal function. Such a function would not require that *swallow* protein be associated with *bicoid* message or any other RNA, so demonstration that the putative RNA-binding motif in *swallow* actually functions as such *in vivo* would require a rethinking of the *bicoid* message localization process. We expect to obtain considerable information about *bicoid* message localization by determining the subcellular location and molecular interactions of the *swallow* protein. The molecular tools for these experiments are now available, so a more complete picture of this interesting process should be available soon.

Acknowledgments

Original work reported here was supported by research Grants DCB 87-02160 from the National Science Foundation and GM 41513 from the National Institutes of Health. N.J.P. was supported by an NIH Training Grant in Genetics and Regulation.

References

Akam, M. (1987). The molecular basis for metameric pattern in the *Drosophila* embryo. *Development (Cambridge, UK)* **101**, 1–22.

Anderson, K. V. (1987). Dorsal–ventral embryonic pattern genes of *Drosophila*. *Trends Genet.* **3**, 91–97.

Bandziulis, R. J., Swanson, M. S., and Dreyfuss, G. (1989). RNA-binding proteins as developmental regulators. *Genes Dev.* **3**, 431–437.

Berleth, T., Burri, M., Thoma, G., Bopp, D., Richstein, S., Frigerio, G., Noll, M., and Nüsslein-Volhard, C. (1988). The role of localization of *bicoid* RNA in organizing the anterior pattern of the *Drosophila* embryo. *EMBO J.* **7**, 1749–1756.

Bohrmann, J., and Gutzeit, H. O. (1987). Evidence against electrophoresis as the principal mode of protein transport in vitellogenic follicles of *Drosophila*. *Development (Cambridge, UK)* **101**, 279–288.

Chao, Y.-C., Donahue, K. M., Pokrywka, N. J., and Stephenson, E. C. (1991). Sequence of *swallow*, a gene required for the localization of *bicoid* message in *Drosophila* eggs. *Dev. Genet.*, in press.

Driever, W., and Nüsslein-Volhard, C. (1988). A gradient of *bicoid* protein in *Drosophila* embryos. *Cell (Cambridge, Mass.)* **54**, 83–93.

Driever, W., and Nüsslein-Volhard, C. (1989). The *bicoid* protein is a positive regulator of *hunchback* transcription in the early *Drosophila* embryo. *Nature (London)* **337**, 138–143.

Frohnhöfer, H. G., and Nüsslein-Volhard, C. (1987). Maternal genes required for anterior localization of *bicoid* activity in the embryo of *Drosophila. Genes Dev.* **1**, 880–890.

Gutzeit, H. O. (1986a). The role of microtubules in the differentiation of ovarian follicles during vitellogenesis in *Drosophila. Roux's Arch. Dev. Biol.* **195**, 173–181.

Gutzeit, H. O. (1986b). The role of microfilaments in cytoplasmic streaming in *Drosophila* follicles. *J. Cell Sci.* **80**, 159–169.

Gutzeit, H. O., and Koppa, R. (1982). Time-lapse film analysis of cytoplasmic streaming during late oogenesis of *Drosophila. J. Embryol. Exp. Morphol.* **67**, 101–111.

Hay, B., Ackerman, L., Barbel, S., Jan, L. Y., and Jan, Y. N. (1988). Identification of a component of *Drosophila* polar granules. *Development (Cambridge, UK)* **103**, 625–640.

King, R. C. (1970). "Ovarian Development in *Drosophila melanogaster.*" Academic Press, New York.

King, R. C., Cassidy, J. D., and Rousset, A. (1982). The formation of clones of interconnected cells during gametogenesis in insects. *In* "Insect Ultrastructure" (R. C. King and H. Akai, eds.), Vol. 1, pp. 3–31. Plenum, New York.

Kouzarides, T., and Ziff, E. (1989). Leucine zippers of fos, jun and GCN4 dictate dimerization specificity and thereby control DNA binding. *Nature (London)* **340**, 568–571.

Macdonald, P. M. (1990). *bicoid* mRNA localization signal: Phylogenetic conservation of function and RNA secondary structure. *Development (Cambridge, UK)* **110**, 161–172.

Macdonald, P. M., and Struhl, G. (1988). Cis-acting sequences responsible for anterior localization of *bicoid* mRNA in *Drosophila* embryos. *Nature (London)* **336**, 595–598.

Mahowald, A. P. (1971). Polar granules of *Drosophila*. III. The continuity of polar granules during the life cycle of *Drosophila. J. Exp. Zool.* **176**, 329-344.

Mahowald, A. P., and Kambysellis, M. P. (1980). Oogenesis. *In* "The Genetics and Biology of *Drosophila*" (M. Ashburner and T. R. F. Wright, eds.), Vol. 2c, pp. 141–224. Academic Press, New York.

Mahowald, A. P., and Strassheim, J. M. (1970). Intercellular migration of centrioles in the germarium of *Drosophila melanogaster. J. Cell Biol.* **45**, 306–320.

Nüsslein-Volhard, C., Frohnhöfer, H. G., and Lehmann, R. (1987). Determination of anteroposterior polarity in the *Drosophila* embryo. *Science* **238**, 1675–1681.

Query, C. C., Bentley, R. C., and Keene, J. D. (1989). A common RNA recognition motif identified within a defined U1 RNA binding domain of the 70k U1 snRNP protein. *Cell (Cambridge, Mass.)* **57**, 89–101.

St. Johnston, D., Driever, W., Berleth, T., Richstein, S., and Nüsslein-Volhard, C. (1989). Multiple steps in the localization of *bicoid* RNA to the anterior pole of the *Drosophila* oocyte. *Development, 1989 Suppl.* 13–19.

Stephenson, E. C., Chao, Y.-C., and Fackenthal, J. D. (1988). Molecular analysis of the *swallow* gene of *Drosophila melanogaster. Genes Dev.* **2**, 1655–1665.

Struhl, G., Struhl, K., and Macdonald, P. M. (1989). The gradient morphogen *bicoid* is a concentration-dependent transcriptional activator. *Cell (Cambridge, Mass.)* **57**, 1259–1273.

Telfer, W. H. (1975). Development and physiology of the oocyte–nurse cell syncytium. *Adv. Insect Physiol.* **11**, 223–319.

Warn, R. M., Gutzeit, H. O., Smith, L., and Warn, A. (1985). F-actin rings are associated with the ring canals of the *Drosophila* egg chamber. *Exp. Cell Res.* **157**, 355–363.

Woodruff, R. I., Kulp, J. H., and LaGaccia, E. D. (1988). Electrically mediated protein movement in *Drosophila* follicles. *Roux's Arch. Dev. Biol.* **197**, 231–238.

Zalokar, M., Audit, C., and Erk, I. (1975). Developmental defects of female-sterile mutants of *Drosophila melanogaster. Dev. Biol.* **47**, 419–432.

3

Actin and Actin-Associated Proteins in *Xenopus* Eggs and Early Embryos: Contribution to Cytoarchitecture and Gastrulation

Elaine L. Bearer
Division of Biology and Medicine
Brown University
Providence, Rhode Island 02912

I. Introduction

Actin is intimately involved in several of the events on which development depends. Some of these events are known to occur in embryos of many widely divergent species. The amphibian embryo has been widely used as a model system for the exploration of cell biological events that contribute to development, in part because it is a vertebrate system that can be fertilized *in vitro*, and is easily manipulated during the early stages of embryogenesis. This article will review the role of the actin-based cytoskeleton of amphibian eggs during the major events of embryogenesis. In addition, some new information on novel actin-binding proteins that may contribute to the structuring of that cytoskeleton will be provided.

Current Topics in Developmental Biology, Vol. 26
Copyright © 1992 by Academic Press, Inc. All rights of reproduction in any form reserved.

II. Oogenesis

It is now apparent from observations in a variety of species that the cytoplasm of the oocyte is highly structured (Bearer, 1991a). In *Drosophila,* maternally synthesized mRNA and protein are asymmetrically distributed, and that distribution is essential for the establishment of the embryonic axes (Nüsslein-Volhard and Roth, 1989). In *Caenorhabditis elegans,* positional information encoding posteriorness is redistributed after fertilization in a process that requires actin (Strome and Hill, 1988), and in ascidian embryos a similar process depends on an actin-dependent cortical contraction that also redistributes cues involved in establishing the posterior parts of the future embryo (reviewed in Jeffrey and Swalla, 1990).

A. The Mitochondrial Cloud

In stage I previtellogenic oocytes of the frog, *Xenopus laevis,* a tight ball of cytoplasm termed the mitochondrial cloud is readily observed in either whole mounts or sections of ovaries (Heasman *et al.,* 1984; see Fig. 4A). The formation of this structure identifies the vegetal hemisphere. In addition to mitochondria, it contains the major cytoskeletal proteins, tubulin, intermediate filaments (Wylie *et al.,* 1985), and actin (T. Whitfield, C. C. Wylie, and E. L. Bearer, unpublished observations), as well as two of the putative factors involved in the future developmental program: the germplasm, containing polar granules, that becomes segregated into the future germ cells during the cleavage stages (Heasman *et al.,* 1984), and Veg-1 mRNA, encoding a member of the transforming growth factor *B* (TGF-*B*) family (Weeks and Melton, 1987; Yisraeli and Melton, 1988).
    Later in oogenesis, the mitochondrial cloud fragments and the same constituents are then found in a shell just within the plasma membrane of the vegetal half of the embryo. Colchicine, which depolymerizes the microtubule system, inhibits the accumulation of the Veg-1 transcript in this shell during stages III and IV of oogenesis, while cytochalasin, which blocks actin polymerization, causes the message to be released from the vegetal shell in stage V or VI oocytes (Yisraeli *et al.,* 1989). It is not yet clear whether this localization of Veg-1 message in the vegetal half is necessary for the future establishment of the embryonic axes. Although the cytochalasin experiments support the notion that actin filaments play a part in retaining the Veg-1 message in its proper location, it is not known how they do this. Other proteins are likely to be involved, either directly as attachment links between the message and the filaments, or indirectly as organizers of the filament system itself. Furthermore, it is not understood why the actin is localized there, whether it is monomeric or polymerized in filaments in this particular place within the cytoplasm (Dent and Klymkowsky, 1988; Franke *et al.,* 1976).

B. Nuclear Actin

In addition to its cytoplasmic roles, actin is also present in the nucleus of the *Xenopus* oocyte (Clark and Merriam, 1977) and may be involved in chromosome condensation, which occurs during oocyte maturation and results in the metaphase II-arrested state of the mature and fertilizable egg (Rungger *et al.*, 1979). Anti-actin antibodies affect transcription of lampbrush chromosomes in oocytes of the urodele, *Pleurodeles watlii* (Sheer *et al.*, 1984), and thus this nuclear actin may also play a role in transcription during oogenesis.

III. Fertilization and Early Cleavage

During fertilization of most species, the sperm entry into the egg cytoplasm, the exocytosis of cortical granules, and the meeting of sperm and egg pronuclei all appear to require actin filaments (Stewart-Savage and Grey, 1982; reviewed in Vaquier, 1981).

A. Cortical Contractions

In *Xenopus*, the pigmented animal hemisphere contracts isometrically within 5–6 min of sperm entry (Elinson, 1975). The contractile cortex of the zygote is 0.5–3 μm thick, excludes yolk platelets, and contains cortical granules, pigment granules, and actin filaments (Merriam *et al.*, 1983; Christensen *et al.*, 1984). Within 20–30 min the center of gravity is shifted to the vegetal pole, so that all fertilized eggs are oriented with the animal side up, while unfertilized eggs orient randomly, often on their sides. The first contraction relaxes and microtubule networks form in the vegetal hemisphere that are necessary for the subsequent rotation, which sets up the gray crescent and specifies the embryonic axes (see Houliston and Elinston, Chapter 4 in this volume). A second contraction of the pigmented cortex occurs approximately 70 min after fertilization, relaxing as the first cleavage furrow forms. Cytochalasin B has been shown to have no effect on these contractions in *Xenopus*, affecting only the formation of the cleavage furrow. However, by electron microscopy it appears that cytochalasin B does not completely depolymerize the actin filaments even when injected into the egg during these contractions (Luchtel *et al.*, 1976). Thus, the developmental significance of these contractions and their dependence on actin have not been defined. That depolymerization of actin filaments may be permissive for cortical rotation is implied by the result that cytochalasin treatment induces formation of a morphological gray crescent in unactivated eggs lying on their sides (Manes *et al.*, 1978).

Overaged, unfertilized eggs will undergo a similar contraction followed by a stretching of the animal pole cortex over the vegetal pole reminiscent of epiboly, the movements of the surface epithelium during gastrulation (E. L. Bearer, personal observations; Holtfreter, 1943). This behavior suggests that the cytoskeletal machinery necessary for this movement is present and preprogrammed in the egg.

While the animal hemisphere is contracting, the actin network of the vegetal pole may depolymerize, because after fertilization Veg-1 message is released from its tight shell in the cortex of the vegetal hemisphere, but remains in the vegetal cytoplasm just as it does in mature oocytes treated with cytochalasin B (Weeks and Melton, 1987). In contrast, the polar granules and germplasm, perhaps retained by vimentin-like intermediate filaments, remain in the cortex of the vegetal pole, aggregate, and are divided equally to the first four daughter cells, and unequally in subsequent divisions until they become segregated into the future germ cells (Heasman et al., 1984).

B. Biochemistry of the Contractile Machinery

The contraction of the animal cortex has been studied in some detail biochemically. Extracts of mature eggs can be induced to form a gel in the presence of $0.6\,M$ sucrose. The gel will contract with calcium, with or without ATP (Clark and Merriam, 1978). It contains a 43-kDa protein (actin) and a 250-kDa protein, as well as 53- and 68-kDa proteins in lower amounts. Only in the absence of exogenous ATP is a 200-kDa putative myosin found associated with the contractile gel. Cortices of eggs and embryos can be dissected off the underlying cytoplasm (Franke et al., 1976), or eggs can be cut in half (Christensen et al., 1984). In either of these preparations, the contraction can be elicited and the necessary components extracted and reconstituted. Cortices from oocytes prior to maturation (prior to attaining metaphase II arrest) are refractile to such contractions. A Xenopus myosin II-like molecule was found to be soluble under these conditions but necessary for the contractions (Christensen et al., 1984).

The animal cap contraction ultimately results in the formation of the first cleavage furrow, which first becomes visible approximately 75 min after fertilization over the animal hemisphere. The subsequent cleavages all depend on actin for both the formation of the cleavage furrow and the proper segregation of cytoplasmic components. In addition, the cells appear to adhere to each other via an actin-based process, because cytochalasin treatment results in the cells becoming dissociated even within an intact vitelline membrane.

IV. Gastrulation

A. Cell Shape Changes and Cell Movements

The microtubule-dependent rotation of the cortex over the underlying cytoplasm results in an asymmetric distribution of unknown factors that defines the dorsal lip of the blastopore and the embryonic axis. Whatever these asymmetrically distributed factors might be, their continued localization in that region of the embryo probably depends both on proper segregation of the factors into daughter cells and on some sort of cytoskeletal stabilization. By midblastula transition when the zygotic genome is activated, during interphase of nuclear cycle 9, the cells are apparently programmed to begin the movements of gastrulation. These movements are symmetrical in embryos that were prevented from rotating by UV irradiation (Malacinski *et al.*, 1977; Gerhart *et al.*, 1989), producing ventralized embryos; they are also inaccurate in embryos in which transcription is blocked, suggesting that maternal cues and maternally deposited proteins as well as newly synthesized zygotic gene products are necessary for gastrulatory movements to occur.

The first cells to change shape are the surface cells of the gray crescent, in the marginal zone at the boundary between the large, yolky cells of the vegetal pole and the smaller animal pole cells (Keller, 1981; Hardin and Keller, 1988). These cells have been named "bottle cells" because of their characteristic shape. Initially, these cells remain vegetal to the involuting marginal zone surface epithelium, but they are ultimately also carried inside and contribute to the anteriormost tip of the archenteron (Keller, 1981). During the next two stages of embryogenesis, a wave of bottle cell formation encircles the equator, spreading laterally from either side of the dorsal lip and meeting at the ventral lip to form the complete ring of the blastopore.

During the complex movements that follow, two events stand out: migration of the deep cells of the marginal zone, and spreading of the surface cells of the animal cap. The deep cells originate in the marginal zone of the dorsal lip of the blastopore, in the region between the lateral floor of the blastocoel cavity and the gray crescent. They ultimately contribute to mesodermal structures. Early in gastrulation, the deep cells of the dorsal lip of the blastopore migrate up over the roof of the blastocoel, apparently on a fibronectin substrate (Winklbauer, 1990, and references therein). These migrations appear to involve complex cell shape changes (Winklbauer, 1990) with protrusions of filopodia rather than the conventional leading edges observed in cultured fibroblasts. In contrast, the surface cells apparently retain some of their cell–cell contacts and move collectively as a sheet, which ultimately spreads out as a single layer over the entire surface of the embryo, engulfing the vegetal mass. Only a narrow band of surface cells at the equator actually involutes and these form the inner surface of the roof of the archenteron. The yolky vegetal cells are apparently passively carried along by these movements.

Further evidence that these movements are a combined result of maternal and zygotic elements has been shown by the behavior of cells taken from different parts of the embryo at different stages of development and observed *in vitro* (Nakatsuji, 1986; Winklbauer, 1988). When cells from the animal cap are plated on fibronectin-coated coverslips, they do not migrate until after stage 8. The inner cells from the equatorial region are the earliest to migrate and move more vigorously than cells from other parts of the embryo. The surface epithelium has different adhesion properties than the endodermal cells of the vegetal pole. Although vegetal pole cells adhere to fibronectin, they do not migrate.

B. Actin-Binding Proteins

Because these movements rely on actin, it becomes particularly interesting to determine what molecules regulate the actin filament network such that it can be at the right place and the right time for stabilization of the Veg-1 message in the vegetal shell, and also be involved in such different concerted cell movements as the shape changes of the surface epithelium and the cell migrations of the deep cells. Some of the factors that regulate actin polymerization should contribute to both processes, while others must be specific, for each process is different.

To date, only a few of the protean numbers of actin-binding proteins have been conclusively found in *Xenopus* oocytes and embryos. These include gelsolin (Ankenbauer *et al.*, 1988), spectrin (Giebelhaus *et al.*, 1987), and myosin (Christensen *et al.*, 1984). Gelsolin is an actin filament-severing and barbed end-binding protein that nucleates filaments from the pointed, or slow-growing, end. In the oocyte, gelsolin apparently binds monomers and prevents the vast pool of actin (4.1 mg/ml) from polymerizing inappropriately (Merriam and Clark, 1978). In addition to these actin-binding proteins, vinculin and talin, cytoskeletal proteins involved in the formation of adhesion plaques, have been found in frog eggs and embryos (Evans *et al.*, 1990), as well as homologs of the integrin (fibronectin) receptors (DeSimone and Hynes, 1988). Talin is known to bind directly to the fibronectin receptor (Horwitz *et al.*, 1986). Thus its contribution to the fibronectin-based migrations of the deep cells over the blastocoel roof may shed light on its involvement in cellular migrations in other systems. That these proteins are present in the egg suggests that they are stockpiled for future use during gastrulation, and that they are not the key elements whose synthesis is triggered by maternal cues in the deep cells.

C. Actin-Associated Proteins Involved in Cell Migrations

It seems likely that key cytoskeletal regulatory elements exist and are turned on in specific cells prior to or during midblastula transition. These key elements would be responsible for initiating the cytoskeletal changes necessary for the different types of cell shape changes that result in the different types of movements of gastrulation. For example, simplisticly, if the cell shape changes involved in epiboly relied on a contractile protein, perhaps calcium channels regulating the action of that protein, or the protein itself, would be synthesized only in sufficient levels to respond to contractile stimuli in the surface cells of the animal cap. Or, if actin polymerization–depolymerization is a necessary part of inner cell migrations, then those proteins involved in such reorganizations of the actin cytoskeleton would be specifically induced in deep cells and not in vegetal cells. Conversely, those actin-binding proteins involved in maintenance of the cell cortex or other housekeeping-type actin filament behavior would be stockpiled in the egg, and not differentially expressed just prior to gastrulation. Alternatively, key proteins regulating these disparate actin behaviors may be stockpiled in the egg, either localized in particular regions consistent with their future contribution or becoming localized in those regions during the rotations, contractions, or early cleavages or embryogenesis.

D. MAb 2E4 Antigen

To test this idea, the author has investigated the distribution of an actin-associated protein recognized by the monoclonal antibody, 2E4 (MAb 2E4), in eggs and early embryos. This antibody was originally raised against proteins extracted from human blood platelets that were retained on an F-actin column and eluted with ATP (Bearer, 1991b). The antibody stains the leading edge of migrating chick embryo fibroblasts, and the antigen–antibody complex associates with the barbed ends of purified actin filaments *in vitro* in an ATP-sensitive manner (Bearer, 1991b,c). Because of its association with the barbed end of actin filaments, which is thought to be the end that elongates through polymerization to form filaments, and because of its location in cells at sites where such polymerization is occurring, the antigen may be involved in regulating actin filament polymerization. The antibody recognizes a 43-kDa protein in blots from a wide variety of species from fruit flies to humans. This band comigrates and copurifies with actin, but can be separated from the bulk of actin in the presence of urea at low pH (Bearer, 1991b).

In polyethylene glycol-embedded sections of stage 9 *Xenopus laevis* embryos, MAb 2E4 stains most prominently the deep cells of the marginal zone (arrow in Fig. 1A). There is some staining of cell margins over the entire animal cap, but the large yolky cells in the vegetal pole are only lightly stained. In contrast, an anti-actin antibody of the same isoform as MAb 2E4, used as a positive control, stains cells equally throughout the embryo (Fig. 1B). In two other embryos viewed at higher magnification (Fig. 1C and D), MAb 2E4 staining again appears more prominent on one side of each of the cells it stains. Both surface epithelium and deep cells in the marginal zone are strongly stained, as well as one side of deep cells in the animal cap. Again, staining of vegetal cells is not observed. A single small cell at the crotch of the blastocoel is also strongly positive (Fig. 1C, arrow).

When the mesoderm is dissected from stage 25 neurulas, cut into small pieces, and cultured on matrigel-coated coverslips, individual cells will crawl out of the mass of tissue. Cells in such cultures were fixed and stained with both MAb 2E4 (Fig. 2A) and fluorescein-conjugated phalloidin, which stains actin filaments (same cell, Fig. 2B). Comparisons of such double-labeled cells reveal that MAb 2E4 antigen appears as punctate dots that overlie the ends of actin filament bundles in the pseudopods projecting from the cell body. Punctate or granular staining by MAb 2E4 is also present in the body of the cell, but the actin filament

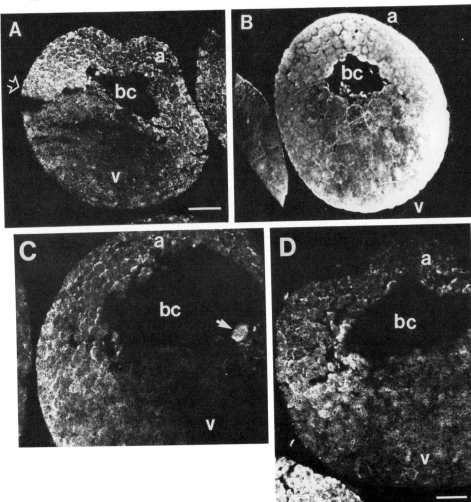

Fig. 1 Polyethylene glycol-embedded sections of stage 8 and 9 *Xenopus* embryos stained with MAb 2E4 or an anti-actin monoclonal antibody according to Wylie *et al.* (1985). (A) MAb 2E4-stained section of a stage 9 embryo. bc, Blastocoel cavity; v, vegetal mass. Arrow indicates bright staining by the antibody in the presumptive dorsal lip. (B) Anti-actin monoclonal antibody stains a stage 8 embryo diffusely, with more concentrated staining at the cortex of individual cells. bc, Blastocoel cavity; a, animal pole; v, vegetal pole. Bar in (A) (= 0.4 mm) gives magnification for (A) and (B). (C and D) Higher magnification of sections of other embryos stained with MAb 2E4. Brightest staining is seen in the cells of the presumptive mesoderm, the deep cells of the marginal zone. The basal edge of cells over the animal cap are also stained. bc, Blastocoel cavity; a, animal cap; v, vegetal mass. Arrow in (C) points out a single small cell, lying on the floor of the blastocoel, that stains strongly. Bar in (D) (0.22 mm) gives magnification for (C) and (D).

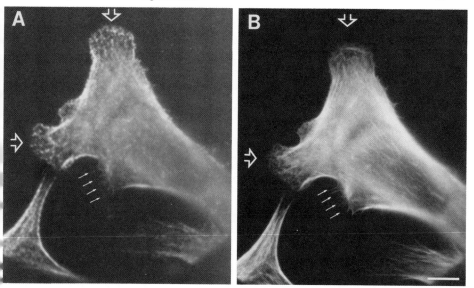

Fig. 2   Mesodermal cells from a stage 25 embryo crawling on matrigel-coated coverslips, fixed in 4% paraformaldehyde and stained by indirect immunofluorescence with MAb 2E4 and phalloidin. (A) MAb 2E4 stains in a punctate pattern both the cytoplasm and the pseudopods (large open arrows). Microspikes or filopodia extending from the lateral cell borders are also weakly stained (small arrows). (B) The same cell also stained with phalloidin displays fine filaments in the pseudopods (open arrows). The ends of these filaments appear to coincide with the points stained by MAb 2E4. Microspikes or filopodia extending from the lateral margin also stain (small arrows). Bar = 25 μm.

staining is too dense for a correlation to be made. Microspikes extending from the side of the cell are also stained by both MAb 2E4 and phalloidin, and a granular staining of MAb 2E4 is observed along bundles of actin filaments in two cells at the lower edge of these micrographs. Because long filament bundles are made up of interdigitating shorter filaments, this punctate staining could reflect the ends of filaments within a bundle.

MAb 2E4 recognizes a single band of apparent M_r 43K in homogenates of human blood platelets and chick embryo fibroblasts (Bearer, 1991b). However, in cytoskeletal extracts of stage 10 or 11 embryos made according to a protocol that extracts the actin-binding protein gelsolin from *Xenopus* oocytes (Ankenbauer *et al.*, 1988), MAb 2E4 blots to five bands of approximate molecular weights of 130K, 95K, 48K, 43K, 25K (Fig. 3A, lane 1: Coomassie-stained gel; lane 2: MAb 2E4-stained Western blot). To see which of these five protein species were responsible for the restricted staining pattern, polyclonal mouse antibodies were made to each band cut from a preparative "curtain" gel (James and Elgin, 1986). These antibodies were then tested by blotting to fresh extract to see whether they recognized bands consistent with the immunogen (Fig. 3B), and by indirect immunofluorescence to see if they recognized the same structures stained by MAb 2E4.

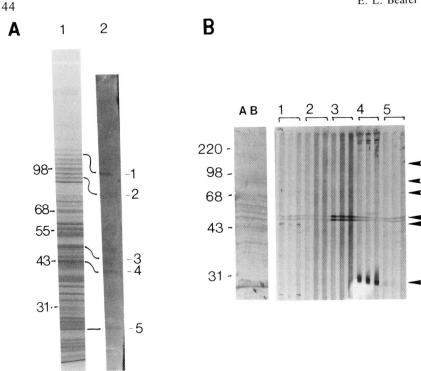

Fig. 3 MAb 2E4 recognizes several bands in blots of cytoskeletal extracts made from stage 10 embryos according to Ankenbauer *et al.* (1988). Lane 1, Coomassie-stained gel; lane 2, Western blot stained with MAb 2E4. Molecular weights for the Coomassie gel are indicated to the left. Amido-black staining of a strip of adjacent nitrocellulose was used to identify bands stained by the antibody. These are indicated by the lines connecting the gels. The bands were numbered and used as immunogens for injection into five separate mice. The identifying numbers are given to the right of lane 2. (B) Western blots of extracts of stage 10 *Xenopus* embryos with antisera resulting from immunizing five mice with individual bands corresponding to those recognized by MAb 2E4 and cut from the preparative gel shown in (A), lane 1. Lane AB: Amido-black staining of transfer blot. Lanes 1: 1/50, 1/100, 1/500 dilutions of antisera resulting from band 1; Lanes 2: 1/50, 1/100, 1/500 dilutions of antisera resulting from band 2; Lanes 3: 1/50, 1/100, 1/500 dilutions of antisera resulting from band 3; Lanes 4: 1/50, 1/100, 1/500 dilutions of antisera resulting from band 4; Lanes 5: 1/50, 1/100, 1/500 dilutions of antisera resulting from band 5. Arrowheads to the right of the blot indicate bands recognized by the antisera. Note that the 75- and 25-kDa bands are stained in lanes 1, 4, and 5, while the 95-kDa band is unique to antiserum 2. Antiserum 3 was so strong that it spread across the blot, obscuring whether the other antisera also cross-react with that band in this blot. In other blots, the other antisera did not appear to recognize that 48-kDa doublet. Molecular weights are indicated to the left.

Two of the antisera recognized bands of the same molecular weight as the immunogen—band 2, 95kDa and band 3, 48kDa doublet (Fig. 3B, lanes 2 and lanes 3). All three other antisera recognized both a 75kDa band and a 25kDa band, as well as another band that was unique to each antiserum, i.e., band 1 (130k immunogen) blots to a 43kDa band (Fig. 3B, lanes 1), band 4 (43kDa immunogen) blots to a >250kDa band (Fig. 3B, lanes 4), and band 5 (25kDa

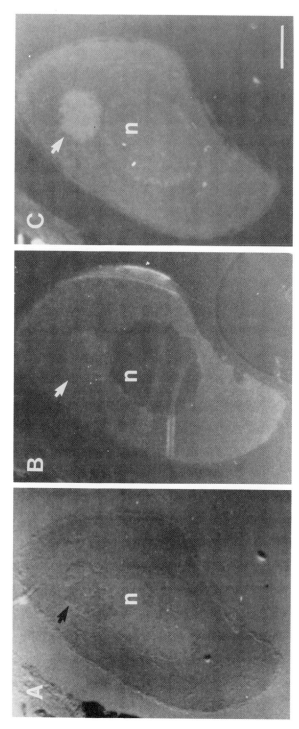

Fig. 4 Plastic-embedded sections of stage 1 oocytes. (A) Nomarski optics identify the mitochondrial cloud (arrow) and the nucleus (n). (B) MAb 2E4 stains the mitochondrial cloud (arrow) and the cytoplasm weakly. The nucleus is not stained. (C) Antiserum 1 (like antisera 4 and 5) stains both the cloud (arrow) and the nucleus. The cytoplasm is also lightly stained. Bar = 25 μm.

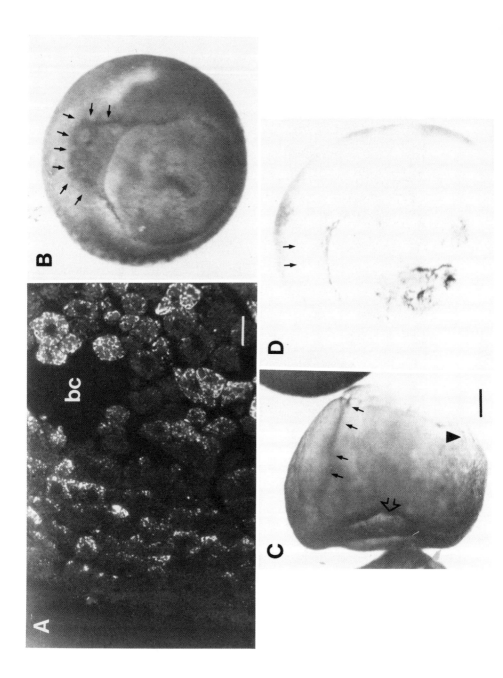

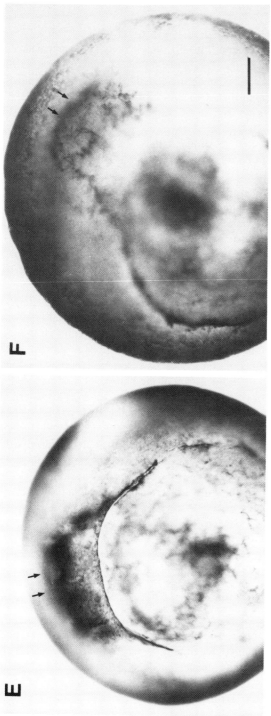

Fig. 5 Staining pattern of antiserum 3 on *Xenopus* embryos. (A) Plastic-embedded section of a stage 9 embryo stained with antiserum 3 indirect immunofluorescence. The angle of the blastocoel cavity (bc) is shown, revealing staining of individual cells of the presumptive mesoderm, and staining of the deep cells over the animal cap. Bar = 0.1 mm. (B–F) Whole mounts of stage 10½ embryos stained with immunoperoxidase method and clarified with butanol/isobutane so that the deep cells of the involuting marginal zone are revealed. (B) Dark reaction product (small arrows) is seen in the involuting marginal zone, perhaps in highest concentration in the bottle cells at the leading margin of involution. (C) A different embryo stained the same way, viewed on its side. A line of reaction product (small arrows) leads up to the floor of the blastocoel (open arrow). The marginal zone around the embryo is stained, while the vegetal cells are not visible because they are not stained, and no ventral lip is apparent yet (arrowhead). (D) A control embryo incubated with secondary antibody only. No reaction product is seen in the involuting marginal zone (small arrows). Bar = 1 mm. (E and F) Higher magnifications of two more embryos, showing similar staining of the involuting cells of the dorsal lip of the blastopore (small arrows). Bar = 0.5 mm.

immunogen) blots to a 130kDa band (Fig. 3B, lanes 5). This complicated blotting pattern suggests that, while the 95kDa and 48kDa species are distinct, the
130kDa, 75kDa, 43kDa, and 25kDa bands are related to each other antigenically.
A 75kDa band was not seen in the MAb 2E4 blot, perhaps because this is a
proteolytic fragment that does not contain the antigenic epitope recognized by the
monoclonal antibody.

The three antisera that blotted to similar bands also gave similar staining
patterns on stage I oocytes. In plastic sections of stage I oocytes, the mitochondrial cloud is readily identified by Nomarski optics (Fig. 4A, arrow), while the
nucleus (Fig. 4A–C, "n") is also recognizable. In serial sections stained with
MAb 2E4, the mitochondrial cloud is more brightly fluorescent than the
cytoplasm, and the nucleus is not stained at all (Fig. 4B). Both nucleus and
mitochondrial cloud are brightly stained by anti-actin antibodies (not shown). In
other serial sections of the same oocyte, antisera 1, 4, and 5 stained the mitochondrial cloud like MAb 2E4, but stained the nucleus as well (antiserum 1 is
shown in Fig. 4C). Antiserum 4 also stained the vitelline membrane during its
formation later in oogenesis. In contrast, antiserum 3 stained small granular
material in the cytoplasm only, and antiserum 2 did not stain oocytes.

The most remarkable staining pattern of the 2E4 monoclonal antibody was the
staining pattern in the late blastula (Fig. 1). This pattern was mimicked by
antiserum 3. By indirect immunofluorescence performed on plastic sections,
antiserum 3 recognized the cells in the crotch of the blastocoel, and one side of
the cells over the animal cap (Fig. 5A). In these preparations, the pattern of
staining was granular. In whole mounts of gastrulating embryos (stage 10)
stained with antiserum 3 by the indirect immunoperoxidase method and clarified
by n-butanol, a dark reaction product is readily observed in the deep cells
beneath the dorsal lip of the blastopore (Fig. 5B). At this time point, bottle cells
are at the leading margin of the involuting tissue (Hardin and Keller, 1988),
which is the point of darkest staining by antiserum 3. In an embryo observed
lying on its side, oriented with the dorsal lip toward the top of the photograph and
the animal cap to the left (Fig. 5C), a line of reaction product is seen (small
arrows) that extends up to the floor of the blastocoel (larger arrow). The region
just vegetal to the blastocoel floor is more darkly stained, and the vegetal pole is
not stained. At this time point, actual involution has extended only about half
way to the roof of the blastocoel, and thus the staining precedes involution. The
roof of the blastocoel has collapsed in this embryo. Bottle cells would not yet
have formed in the ventral lip, and no cell movements would have begun. No
reaction product is seen in this region, either (Fig. 5C, arrowhead). Parallel
staining of embryos from the same preparation, in which the primary antibody
was omitted, have no reaction product detectable deep to the dorsal lip, although
the pigment line can be seen to identify it. A nonspecific reaction product is
present at the vegetal pole (Fig. 5D). Higher magnifications of other embryos at
the same stage of gastrulation reveal the intense staining of the deep cells in the
dorsal lip (Fig. 5E and F).

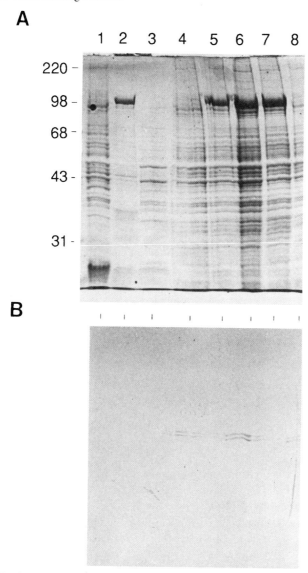

Fig. 6 A developmental profile of the expression in eggs and embryos of the 48-kDa doublet recognized by antiserum 3. (A) Coomassie-stained 10% polyacrylamide SDS gel of cytoskeletal extracts and (B) Western blot with antiserum 3 of Lane 1, ovaries; lane 2, laid eggs; lane 3, 30 min after fertilization; lane 4, stage V embryos; lane 5, stage VI embryos; lane 6, stage 7 embryos; lane 7, stage 8 or 9 embryos; lane 8, stage 10 embryos. Five embryos were extracted for each lane. Molecular weights are given to the left.

To study the expression pattern of the 48-kDa doublet recognized by antiserum 3, extracts were made of eggs and embryos, the proteins separated by poly-acrylamide gel electrophoresis, and the gel either stained with Coomassie to visualize the protein bands (Fig. 6A) or transferred to nitrocellulose and blotted with antiserum 3. Five eggs or embryos were used for each lane. The 48-kDa antigen doublet was not detected until stage 5 of embryogenesis. Both bands of the doublet were equally intense relative to each other at each stage. It was found that some protein is also present in the oocyte, but it is not extracted under the conditions that extract it from the embryo, and is detected only in the pellet.

V. Conclusion

This chapter has reviewed the contribution of actin to the events of oogenesis and early embryogenesis in eggs and embryos of the frog X. laevis. This contribution includes the organization of the mitochondrial cloud and the maintenance of the vegetal shell of cytoskeletal components that limits the diffusion of Veg-1 mRNA during oogenesis, the cortical contractions that occur following fertilization, the localization of as yet undefined developmental cues in appropriate places within the cytoplasm and their appropriate apportionment to daughter cells during cytokinesis, and the cell shape changes and motility that produce gastrulation.

In addition, preliminary results have been presented showing that a group of proteins antigenically related to an actin-associated antigen, the MAb 2E4 anti-gen, is localized in regions of the egg and early embryo where actin filaments play an incisive role in these events. Finally, it has been shown that at least one of these proteins, the 48-kDa doublet recognized by antiserum 3, is preferentially expressed prior to midblastula transition in the subpopulation of cells that begin the movements of gastrulation—the deep cells of the dorsal lip of the blastopore.

Many more questions remain to be answered, both about these antigens and about the role of actin itself, in the formation and movements of the components of the mitochondrial cloud during oogenesis, and the regulation of cell motility on which gastrulation depends. For example, does actin depolymerize in the vegetal pole on fertilization? What effect does the apical contraction, occurring after fertilization, have on the subsequent rotation that determines the dorsal lip and the anterior–posterior axis of the embryo? Are there key regulatory proteins that mediate cell motility? Is their synthesis one of the triggers for subsequent cell movements? If so, how does their presence or activity affect actin filament dynamics in the cell?

Acknowledgments

I am indebted to Christopher Wylie for my introduction to Xenopus embryos, and to him and Tania Whitfield for the preliminary staining of embryos and eggs with MAb 2E4. I also wish to thank Ray

Keller for advice and discussions. This work is supported by a grant from the American Cancer Society, California chapter (#SS44-89), and the Frederick Bang Award, Marine Biology Lab, Woods Hole, Massachusetts.

References

Ankenbauer, T., Kleinschmidt, J. A., Vanderkerckhove, J., and Franke, W. W. (1988). Proteins regulating actin assembly in oogenesis and early embryogenesis of *Xenopus laevis:* Gelsolin is the major cytoplasmic actin-binding protein. *J. Cell Biol.* **107**, 1489–1498.

Bearer, E. L. (1991a). Actin in *Drosophila* embryos: Is there a relationship to positional cue localization? *BioEssays* **13**(4), 199–204.

Bearer, E. L. (1991b). A 43 kD protein associated with the leading edge of migrating cells. Submitted for publication.

Bearer, E. L. (1991c). Direct observations of actin filament severing by gelsolin, and binding by gCap 39 and CapZ. *J. Cell Biol.* (in press).

Christensen, K., Sauterer, R., and Merriam, R. W. (1984). Role of soluble myosin in cortical contractions of *Xenopus* eggs. *Nature (London)* **310**, 150–151.

Clark, T. G., and Merriam, R. W. (1977). Diffusible and bound actin in nuclei of *Xenopus laevis* oocytes. *Cell (Cambridge, Mass.)* **12**, 883–891.

Clark, T. G., and Merriam, R. W. (1978). Actin in *Xenopus* oocytes. I. Polymerization and gelation *in vitro. J. Cell Biol.* **77**, 427–447.

Dent, J. A., and Klymkowsky, M. W. (1988). Wholemount analysis of cytoskeletal reorganization and function during oogenesis and early embryogenesis in *Xenopus. In* "Cell Biology of Fertilization" (H. Shatten and G. Shatten, eds.). Academic Press, Orlando, Florida.

DeSimone, D. W., and Hynes, R. O. (1988). *Xenopus laevis* integrins: Structural conservation and evolutionary divergence of integrin B subunits. *J. Biol. Chem.* **263**, 5333–5340.

Elinson, R. P. (1975). Site of sperm entry and cortical contraction associated with egg activation in the frog *Rana pipiens. Dev. Biol.* **47**, 257–268.

Evans, J. P., Page, B. D., and Kay, B. K. (1990). Talin and vinculin in the oocytes, eggs and early embryos of *Xenopus laevis:* A developmentally regulated change in distribution. *Dev. Biol.* **137**, 403–413.

Franke, W. W., Rathke, P. C., Seib, E., Trendelenburg, M. F., Osborn, M., and Weber, K. (1976). Distribution and mode of arrangement of microfilamentous structures and actin in the cortex of the amphibian oocyte. *Cytobiologie* **14**, 111–130.

Gerhart, J., and Keller, R. (1986). Region-specific cell activities in amphibian gastrulation. *Annu. Rev. Cell Biol.* **2**, 206–229.

Gerhart, J., Danilchik, M., Doniach, T., Roberts, S., Rowning, B., and Stewart, R. (1989). Cortical rotation of the *Xenopus* egg: Consequences for the anteroposterior pattern of embryonic development. *Development, 1989 Suppl.* 37–51.

Giebelhaus, D. H., Zelus, B. D., Henchman, S. K., and Moon, R. T. (1987). Changes in the expression of a-Fodrin during embryonic development of *Xenopus laevis. J. Cell Biol.* **105**, 843–853.

Hardin, J., and Keller, R. (1988). The behavior and function of bottle cells during gastrulation of *Xenopus laevis. Development (Cambridge, UK)* **103**, 211–230.

Heasman, J., Quarmby, J., and Wylie, C. C. (1984). The mitochondrial cloud of *Xenopus* oocytes: The course of germline granule material. *Dev. Biol.* **105**, 458–469.

Holtfreter, J. (1943). Properties and functions of the surface coat in amphibian embryos. *J. Exp. Zool.* **93**, 251–323.

Horwitz, A., Duggan, K., Buck, C., Beckerle, M. C., and Burridge, K. (1986). Interaction of plasma membrane fibronectin receptor with talin, a transmembrane linkage. *Nature (London)* **320**, 531–533.

James, T. C., and Elgin, S. (1986). Identification of a nonhistone chromosomal protein associated with heterochromatin in *Drosophila melanogaster* and its gene. *Mol. Cell. Biol.* **6**(11), 3862–3872.

Jeffrey, W. R., and Swalla, B. J. (1990). The myoplasm of ascidian eggs: A localized cytoskeletal domain with multiple roles in embryonic development. *Semin. Cell Biol.* **1**, 373–381.

Keller, R. E. (1981). An experimental analysis of the role of bottle cells and the deep marginal zone in gastrulation of *Xenopus laevis*. *J. Exp. Zool.* **216**, 81–101.

Keller, R. E. (1984). The cellular basis of gastrulation in *Xenopus laevis:* Active, postinvolution convergence and extension by mediolateral interdigitation. *Am. Zool.* **24**, 589–603.

Luchtel, D., Bluemink, J. G., and De Laat, S. W. (1976). The effect of injected cytochalasin B on filament organization in the cleaving egg of *Xenopus laevis*. *J. Ultrastruct. Res.* **54**, 406–419.

Malacinski, G. M., Brothers, A. J., and Chung, H. M. (1977). Destruction of the neural induction system of the amphibian egg with ultraviolet radiation. *Dev. Biol.* **56**, 24–39.

Manes, M. E., Elinson, R. D., and Barbieri, F. D. (1978). Formation of the amphibian grey crescent: Effects of colchicine and cytochalasin B. *Wilhelm Roux's Arch. Dev. Biol.* **185**, 99–104.

Merriam, R. W., and Clark, T. G. (1978). Actin in *Xenopus* oocytes II. Intracellular distribution and polymerizability. *J. Cell Biol.* **77**, 439–447.

Merriam, R. W., Sauterer, R. A., and Christensen, K. (1983). A subcortical, pigment-containing structure in *Xenopus* eggs with contractile properties. *Dev. Biol.* **95**, 439–446.

Nakatsuji, N. (1986). Presumptive mesodermal cells from *Xenopus laevis* gastrulae attach to and migrate on substrata coated with fibronectin or laminin. *J. Cell Sci.* **86**, 109–118.

Nüsslein-Volhard, C., and Roth, S. (1989). Axis determination in insect embryos. *Ciba Found. Symp.* **144**, 37–55, discussion 55–64, 92–98.

Rungger, D., Rungger-Braendle, E., Chaponnier, C., and Gabbiani, G. (1979). Intranuclear injection of anti-actin antibodies into *Xenopus* oocytes blocks chromosome condensation. *Nature (London)* **282**, 320–321.

Sheer, U., Hinssen, H., Franke, W. W., and Jockusch, B. M. (1984). Microinjection of actin-binding proteins and actin antibodies demonstrates involvement of nuclear actin in transcription of lampbrush chromosomes. *Cell (Cambridge, Mass.)* **39**, 111–122.

Stewart-Savage, J., and Grey, R. D. (1982). The temporal and spatial relationships between cortical contractions, sperm tail formation, and pronuclear migration in fertilized *Xenopus* eggs. *Wilhelm Roux's Arch. Dev. Biol.* **191**, 241–245.

Strome, S., and Hill, D. P. (1988). Early embryogenesis in *C. elegans:* The cytoskeletal and spatial organization of the zygote. *BioEssays* **124**, 1–8.

Vacquier, V. D. (1981). Dynamic changes of the egg cortex. *Dev. Biol.* **84**, 1–26.

Weeks, D. L., and Melton, D. A. (1987). A maternal mRNA localized to the vegetal hemisphere in *Xenopus* eggs codes for a growth factor related to TGF-B. *Cell (Cambridge, Mass.)* **51**, 861–867.

Winklbauer, R. (1988). Differential interaction of *Xenopus* embryonic cells with fibronectin *in vitro*. *Dev. Biol.* **130**, 175–183.

Winklbauer, R. (1990). Mesodermal cell migration during *Xenopus* gastrulation. *Dev. Biol.* **142**, 155–168.

Wylie, C. C., Brown, D., Godsave, S. F., Quarmby, J., and Heasman, J. (1985). The cytoskeleton of *Xenopus* oocytes and its role in development. *J. Embryol. Exp. Morphol.* **89**, Suppl., 1–15.

Yisraeli, J. K., and Melton, D. A. (1988). The maternal mRNA Vg1 is correctly localized following injection into *Xenopus* oocytes. *Nature (London)* **336**, 592–595.

Yisraeli, J. K., Sokol, S., and Melton, D. A. (1989). The process of localizing a maternal messenger RNA in *Xenopus* oocytes. *Development, 1989 Suppl.* 31–36.

4

Microtubules and Cytoplasmic Reorganization in the Frog Egg

Evelyn Houliston and Richard P. Elinson
Department of Zoology
University of Toronto
Toronto, Ontario M5S 1A1, Canada

I. Introduction
II. Microtubule Dynamics in Frog Eggs
   A. Microtubule Behavior and Cell Cycle Progression
   B. Microtubule-Organizing Centers
   C. Microtubule-Associated Proteins and Motors
III. Localizing Cytoplasm in the Unfertilized Egg
   A. Inside–Outside Polarity, Animal–Vegetal Polarity
   B. Microtubules during Oogenesis
   C. Microtubules and Oocyte Maturation
IV. Reorganizing Cytoplasm in the Fertilized Egg
   A. Cortical Rotation (Gray Crescent Formation)
   B. The Vegetal Microtubule Array
   C. Possible Mechanisms of Force Generation
V. Conclusions
   References

I. Introduction

It is convenient to consider the fertilized egg as the starting point for animal development. Packaged in a single cell is a full set of DNA and the cytoplasmic context to enable the complex organization of the adult to develop. At some early stage of embryogenesis, asymmetries in organization must be established as a framework on which the rest is elaborated. This means specifying the main body axes: dorsal–ventral and anterior–posterior, as well as the basic "germ layer" differences that can be considered as a third, "outside–inside" (ectoderm, meso-derm, endoderm) axis. The frog egg achieves this in two stages. Some asymme-tries of cytoplasmic organization are already present in the unfertilized egg. These are sufficient to guide the main cell movements of gastrulation and thus the establishment of inside–outside, and to some degree anterior–posterior, dif-ferences. Rearrangement of the egg cytoplasm during the first cell cycle follow-ing fertilization then creates further differences. These are essential for the cre-ation of a specialized "organizer" region on one side of the embryo, necessary

Current Topics in Developmental Biology, Vol. 26
Copyright © 1992 by Academic Press, Inc. All rights of reproduction in any form reserved.
53

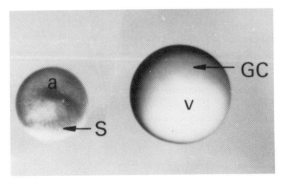

Fig. 1   Eggs from *Xenopus laevis* (left, diameter 1.2 mm) and *Rana pipiens* (right, diameter 1.8 mm) fixed toward the end of the first cell cycle, after cortical rotation has completed embryonic axis specification. They are viewed obliquely such that the animal pole (a) is visible in the *Xenopus* egg and the vegetal pole (v) in the *Rana* egg. The direction of rotation, and thus the orientation of the future dorsal ventral axis of the embryo, is clear in *Rana* eggs from the position of the gray crescent (GC) and can be inferred in some *Xenopus* eggs by swirls in the animal pigment (S).

for the development of the dorsal and anterior structures (see Elinson and Kao, 1989, for review). Thus, by the time of first cleavage, cytoplasmic asymmetries specifying all the embryonic body axes are set up. The future orientation of these axes can be deduced by examining the pigment patterns on the egg (Fig. 1).

In this article we will discuss the establishment of organization in the frog egg cytoplasm in the unfertilized egg, and its subsequent reorganization during the first cell cycle, in the hope that examination of these processes may reveal the nature of developmentally important differences and the mechanisms by which they are established. We will concentrate on the role of microtubules in these events. Descriptions of microtubule redistribution that correlate with cytoplasmic reorganization, and the disruption of cytoplasmic reorganization by agents that affect microtubule stability, have implicated microtubules in both steps of organization.

As well as being an interesting cell because of its developmental potential, the frog egg is also a very rapidly dividing cell. Its cytoplasm and distribution of microtubules undergo repeated remodeling at each transition between interphase and mitosis/meiosis. Before examining aspects of microtubule behavior associated with developmental events, we need to consider microtubule changes related to cell cycle progression, in order to try to dissociate the two.

II. Microtubule Dynamics in Frog Eggs

The frog egg has been exploited as a model system for the examination of microtubule behavior in relation to the cell cycle because fertilization triggers many rounds of rapid and synchronous cell division cycles. We have attempted to

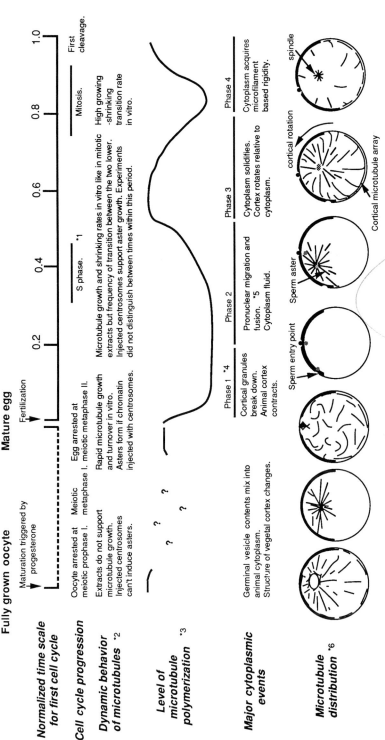

Fig. 2 Summary of microtubule behavior during oocyte maturation and the first cell cycle in relation to cell cycle and developmental events. Sources: *1, Ubbels et al. (1983); *2, Karsenti et al. (1984), Heidemann and Kirschner (1978), Gard and Kirschner (1987a), Verde et al. (1990), Belmont et al. (1990); *3, Elinson (1985), Jessus et al. (1987); *4, Elinson (1983); *5, Stewart-Savage and Grey (1982); *6, Paleček et al. (1985); Gard (1991), Jessus et al. (1986), Elinson and Rowning (1988), Houliston and Elinson (1991a).

draw together the data on microtubule behavior in Fig. 2. The wide variety of different approaches used to examine microtubules in the egg and the different times and events addressed have resulted in some apparent contradictions (see Elinson and Houliston, 1990, for a detailed discussion of these), but general points emerge.

A. Microtubule Behavior and Cell Cycle Progression

Microtubule behavior in dividing cells alternates between interphase and mitotic states (see Brinkley et al., 1976; Kirschner and Mitchison, 1986). In interphase, cytoplasmic conditions favor microtubule growth, so that long microtubules can form without necessarily being nucleated by centrosomes. As the cell enters mitosis conditions change so that cytoplasmic microtubules depolymerize and only centrosome-nucleated microtubules remain. Presumably these changes are important for the formation of the mitotic spindle. In general, the level of tubulin in polymeric form during the meiotic and early mitotic cycles of the frog egg reflects these cell cycle changes, dropping at metaphase and rising in interphase (Fig. 2). The assembly of microtubules during interphase is probably accelerated by a protein that encourages growth at microtubule plus ends (Gard and Kirschner, 1987b), and their rapid disassembly at each mitosis is aided by a microtubule-severing protein that is activated specifically at this time (Vale, 1991).

Not all aspects of microtubule behavior in the frog egg conform to the expected cell cycle pattern. In the unfertilized egg, which is arrested at metaphase II of meiosis, the level of polymer is high (Elinson, 1985; Jessus et al., 1987), and on fertilization (entry to interphase) it drops precipitously. The presence of definitive microtubules in the unfertilized egg, and their absence shortly following fertilization or artificial activation, have been confirmed by electron microscopy (Elinson, 1985) and anti-tubulin staining (Huchon et al., 1988; Houliston and Elinson, 1991a). Despite this atypical presence of cytoplasmic microtubules, dynamic behavior typical of metaphase cells is observed in cytoplasmic extracts of unfertilized eggs made by high-speed centrifugation (Verde et al., 1990). This indicates that "typical" cell cycle behavior can be separated out for study in vitro. Other studies on cell cycle-related microtubule behavior (e.g., Murray and Kirschner, 1989; Belmont et al., 1990; Vale, 1991) have avoided the problem of the unfertilized egg by preparing extracts from activated eggs and then driving them into mitosis in vitro. Appropriate manipulation of conditions can stabilize the interphase or mitotic state.

Although microtubules are present throughout the egg cytoplasm at meiotic metaphase II, there is an apparent depletion in the immediate vicinity of the meiotic spindle (Houliston and Elinson, 1991a). It has been shown by microinjection experiments into eggs that condensed chromatin stimulates the local growth of microtubules on centrosomes and organizes them into a spindle (Karsenti *et al.*, 1984). The restriction of the loss of cytoplasmic microtubules to a small region indicates this influence of the chromatin is quite short ranged. Similar local effects of chromatin on microtubule organization have been demonstrated in the mouse egg (see Maro *et al.*, 1986).

The rapid loss of microtubules associated with the transition to anaphase II is probably due to the sharp rise in the level of cytoplasmic calcium associated with activation (Busa and Nuccitelli, 1985; Kubota *et al.*, 1987). One possible consequence of this high level of cytoplasmic calcium early in the cell cycle is that the first cell cycle is prolonged with respect to the following rapid cleavage divisions. The usual rise in the level of polymerized tubulin on entry into interphase (Elinson, 1985) and the concomitant firming of the cytoplasm (Elinson, 1983) are delayed. This insertion of an additional "phase" of cytoplasmic fluidity in which free microtubules cannot grow may serve a useful purpose by allowing controlled growth of a microtubule aster on the sperm centriole. Growth of the sperm aster seems to be important in moving the pronuclei to meet in the center of the egg (Stewart-Savage and Grey, 1982). Force exerted by the interaction of astral microtubules with the egg surface may tend to push the male pronucleus inward until the aster has grown and migrated to such an extent that this force becomes evenly distributed. A similar mechanism has been proposed to account for the positioning of astral mitotic spindles (Bjerknes, 1986). The sperm aster is probably also instrumental in cueing the later cortical rotation that specifies dorsal–ventral polarity (Manes and Barbieri, 1977; Houliston and Elinson, 1991a; see Section IV,B). Agents that cause precocious microtubule polymerization disrupt both of these processes (Briedis and Elinson, 1982; Scharf *et al.*, 1989; Sakai, 1990).

Many aspects of the cell cycle in early frog development can proceed independently of microtubules (Hara *et al.*, 1980; Gerhart *et al.*, 1984; Clute, 1989; but see Wasserman and Smith, 1978, for conflicting data). In contrast to many other cell types examined, microtubule-disrupting agents do not cause frog blastomeres to arrest in metaphase. Instead, in the absence of spindle microtubules the period spent in the mitotic state, with condensed chromatin, is prolonged (Gerhart *et al.*, 1984; Clute, 1989), as is the case in the sea urchin embryo (Sluder, 1979). This may reflect loss or modification of feedback control mechanisms that usually ensure the correct sequence of mitotic and cleavage events. Some such controls do persist since astral microtubules are needed for positioning and triggering the cleavage furrows (see Elinson, 1990). We can speculate that the loosening of the tight coupling of microtubule reorganization and division cycle may allow microtubules to participate in other events.

B. Microtubule-Organizing Centers

It is generally considered that the metaphase spindle in the unfertilized egg has no centrioles (e.g., see Karsenti *et al.*, 1984). Loss of the centriole present in the early oocyte (Al-Mukhtar and Webb, 1971) is indicated by the relatively poor ability of material associated with the female pronucleus to support microtubule growth during the first cell cycle (Houliston and Elinson, 1991a) and by the lack of a discrete focus for microtubule polymerization in oocytes in regrowth experiments (Gard, 1991). It is clear, however, that material capable of acting as a focus for microtubule polymerization does exist in the full-grown oocyte and egg. In the oocyte, radially arranged microtubules run between the germinal vesicle and the cell periphery (Paleček et al., 1985; Gard, 1991) and will regrow from a region around the germinal vesicle after cold-induced microtubule depolymerization (Gard, 1991). During oocyte maturation, components derived from the germinal vesicle facilitate aster formation in the oocyte cytoplasm (Heidemann and Kirschner, 1978), with microtubule-nucleating activity concentrated around the base of the germinal vesicle (Jessus et al., 1986). Microtubules also polymerize in a radial manner in artificially activated eggs (Manes and Barbieri, 1977), even in the absence of all structures derived from the metaphase II spindle as well as the sperm centriole (Houliston and Elinson, 1991a). Perhaps the yolk-free region in the animal cytoplasm, thought to derive from the germinal vesicle (Ubbels et al., 1983), acts as a store of nucleating material for the cleavage stages of development. Gard et al. (1990) showed that the egg contains sufficient material to create large numbers of centrioles.

The events at the sperm centriole during the initial phase of cytoplasmic fluidity of the first cell cycle remain mysterious. Huge number of microtubules emerge from its vicinity, as the sperm aster grows to fill the animal hemisphere and penetrate the vegetal cytoplasm. It seems impossible that all the astral microtubules could remain joined to the sperm centrosome that nucleates them. Belmont et al. (1990) observed microtubules "releasing" from centrosomes in egg extracts with the aid of minus end-directed microtubule-translocating activity in the cytoplasm. Perhaps such a process could release microtubules from the growing aster. An extension of this idea might be that such microtubule-translocating activity causes sliding between released microtubules and ones anchored to the centrosome, thereby moving some microtubules toward the edge of the expanding aster. An analogous mechanism has been proposed to account for microtubule translocation along axons in slow axonal transport (Vallee and Bloom, 1991). It is also possible that cytoplasmic material capable of nucleating microtubules is added to the centrosome at the heart of the aster as it grows.

C. Microtubule-Associated Proteins and Motors

Several aspects of microtubule behavior in frog eggs are unusual, including the inability of tubulin from full-grown prophase I oocytes to polymerize (Gard and Kirschner, 1987a), the presence of microtubules in metaphase II eggs, the large size of the sperm aster, and the alignment of microtubules over the vast area of the vegetal cortex in the fertilized egg (Elinson and Rowning, 1988). One way to explain these has been to invoke the involvement of specialized proteins that modulate microtubule function or promote their interactions with other elements. Some progress has been made toward the identification of frog egg microtubule-associated proteins (MAPs) by Jessus *et al.* (1985) and Gard and Kirschner (1987b). The MAP characterized by Gard and Kirschner was a specific promoter of microtubule assembly at the plus end.

There is also a body of evidence for the presence of microtubule-associated translocating molecules in frog egg cytoplasm. These molecules are a family of proteins that hydrolyze nucleotide triphosphates to produce microtubule-mediated movement (see Vallee and Shpetner, 1990). The presence of motors such as dynein that create movement toward microtubule minus ends in the egg can be inferred from the movement of pigment granules (Kobayakawa, 1988) and the behavior of microtubules in cytoplasmic extracts (Belmont *et al.*, 1990; Allen and Vale, 1991). Comparison of the movement of vesicles and membrane tubules in interphase and metaphase extracts has further suggested that the activity of both minus and plus end-directed motors is promoted specifically in interphase (Allen and Vale, 1991). *Xenopus* egg kinesin(s) (usually for movement toward plus ends) have been identified by immunoblotting samples enriched in microtubules with antibodies raised against kinesin from other species (Neighbors *et al.*, 1988; Houliston and Elinson, 1991b). Evidence concerning kinesin localization will be discussed later (Section IV,C).

III. Localizing Cytoplasm in the Unfertilized Egg

A. Inside–Outside Polarity, Animal–Vegetal Polarity

The components of the frog egg are commonly classified as being cortical or cytoplasmic. The distinction between the two is determined by microscopic examination, centrifugation experiments, and physical dissection of the eggs (see Elinson, 1980). It should be noted that the definition of the cortex may vary depending on the type of study, and it may change with the progression of developmental time as the cortical components rearrange. All three major cytoskeletal elements show distinct organization in the cortical region in the egg or oocyte, as do membranous organelles of various types, and other proteins and RNAs (see Elinson and Houliston, 1990).

It is tempting to speculate that the cortical location of various components is of relevance to the development later of the outer tissues of the embryo. Byers and Armstrong (1986) showed that very little redistribution of egg surface proteins occurs during cleavage divisions, allowing the development of polarized membrane organization in the superficial epithelial cells of the blastula. Similarly, Wylie *et al.* (1985) suggested that the cytokeratin filament layer in the egg cortex constitutes a localization of this protein for use in the blastula epithelium. The cells that inherit the egg cortex at the blastula stage have distinct behavior, mechanical properties, and differentiation capacities, going on to contribute to epithelia (epidermis and archenteron roof) and neural tissue (see Keller, 1986). Whether any vestige of their inheritance from the egg carries through remains to be established.

The frog egg has an obvious animal–vegetal polarity, with dense pigment granules marking the animal cortex (Fig. 1). Although the pigment boundary is sharp, its precise position is probably insignificant because it varies between frog species, for example, being higher in *Xenopus* than *Rana*. Other cortical elements are also organized asymmetrically along the animal–vegetal axis. For instance, in full-grown oocytes the organization of the cytokeratin filament system is polarized (Klymkowsky *et al.*, 1987), and germ plasm (Heasman *et al.*, 1984) and certain mRNAs (Melton, 1987) are localized preferentially to the vegetal cortex. These three elements become reorganized during meiotic maturation, contributing to the animal–vegetal polarity of the inner cytoplasm. The organization of the cytoplasm is polarized in other respects, too, with distinct differences in such features as the size and density of yolk platelets and density of pigment granules, and the localization to the animal hemisphere of a number of components that derive from the germinal vesicle of the oocyte. The data concerning the development of animal–vegetal polarity have been comprehensively reviewed recently (Dent and Klymkowsky, 1989; Elinson and Houliston, 1990), so we will describe only the main points here.

B. Microtubules during Oogenesis

The position of the mitochondrial cloud at the cortex early in oogenesis defines the vegetal pole (Heasman *et al.*, 1984). This early indicator of polarity is sensitive to microtubule depolymerization, coming adrift from its position in the presence of microtubule poisons (Wylie *et al.*, 1985). Antibody staining of early oocytes reveals a poorly ordered cytoplasmic microtubule network, along with concentrations of microtubules in the mitochondrial cloud itself, in perinuclear aggregates, and in the egg cortex (Gard, 1991). At later stages of oogenesis, tubulin is generally found concentrated in the animal cytoplasm. A fibrous microtubule array is organized in a radial manner around the germinal vesicle, while in the vegetal cytoplasm microtubules are much sparser and less ordered (Paleček *et al.*, 1985; Dent and Klymkowsky, 1989; Yisraeli *et al.*, 1990; Gard, 1991). It appears that this polarized distribution of microtubules in the oocyte contributes to the overall polarity of the cytoplasm, because its disruption interferes with the directed translocation of Vg1 mRNA (Yisraeli *et al.*, 1990). Microtubules are also present in the oocyte cortex (Dumont and Wallace, 1972; Jessus *et al.*, 1986; Huchon *et al.*, 1988; Gard, 1991), and inhibitor studies indicate that they may play a part both in the normal process of yolk uptake (Dumont and Wallace, 1972) and in maintaining the position and structure of a distinct "subcortical" layer (Colman et al., 1981).

C. Microtubules and Oocyte Maturation

The radial and perinuclear microtubules of the full-grown oocyte help to maintain the position of the germinal vesicle deep in the animal cytoplasm (Colman et al., 1981; Lessman, 1987). Although the microtubules reorganize extensively during oocyte maturation (Jessus et al., 1986) they are not critically involved in germinal vesicle breakdown, because this can occur in the presence of microtubule-depolymerizing agents (Lessman, 1987). Whether the meiotic divisions themselves are dependent on microtubules has not been tested.

IV. Reorganizing Cytoplasm in the Fertilized Egg

Two important microtubule-dependent events occur early in frog development. One is the aggregation and ingression of germ plasm at the vegetal pole, a process that occurs over the first few division cycles (Ressom and Dixon, 1988). Microtubules are involved with the initial aggregation of germ plasm patches and later in the "trapping" of the aggregates, but not with their ingression. The second process, gray crescent formation, is of central importance in the establishment of the embryonic body axes. This has attracted much greater attention and so will be discussed here in detail.

A. Cortical Rotation (Gray Crescent Formation)

During the second half of the first cell cycle, the egg cortex undergoes a rotation about a horizontal axis of about 30°. This process results in the formation of a gray crescent in the eggs of certain species, such as Rana pipiens, as part of the clear vegetal cortex is brought to overlay pigmented animal cytoplasm (Fig. 1). It was originally termed the rotation of symmetrization by Ancel and Vintemberger (1948), and is now generally referred to as the cortical rotation (e.g., see Gerhart et al., 1989). Its importance lies in the fact that the rotation is necessary for the localization of dorsal information to a single region of the egg margin. The gray crescent marks the future site of the dorsal lip of the blastopore and the organizer of the amphibian embryo (see Elinson and Kao, 1989, for review).

The direction of the cortical rotation, and thus of the dorsal–ventral axis, is not predetermined in the unfertilized egg, and the gray crescent is able to form at any position around the pigment boundary. In normal conditions, the gray crescent forms roughly opposite the entrance point of the sperm (see Clavert, 1962, for a review of the early literature), and the sperm aster is implicated in directing the rotation (Manes and Barbieri, 1977). It is clear that the sperm aster is not itself critically required for reorganizing the egg cytoplasm. Normal axial development can proceed in activated eggs lacking a sperm aster (Satoh and Shinagawa, 1990). Also, the direction of cortical rotation is a much more accurate indicator of the future position of the dorsal–ventral axis than sperm entry point or animal cytoplasmic asymmetries (Ubbels et al., 1983; Vincent et al., 1986; Vincent and Gerhart, 1987).

A variety of studies have suggested that the machinery for rotation lies close to the surface in the vegetal half of the egg (Malacinski et al., 1975; Manes and Elinson, 1980; Scharf and Gerhart, 1980; Vincent et al., 1987), and that microtubules, but not microfilaments, are actively involved (Manes et al., 1978; Scharf and Gerhart, 1983; Vincent et al., 1987). It should be noted that the rotation machinery itself is not necessary for correct development of the dorsal–ventral axis. Disruptions to it can be overridden by tilting eggs with respect to gravity and thus artificially relocating cytoplasmic constituents (Scharf and Gerhart, 1980; Zisckind and Elinson, 1990). Conversely, dorsal development can be inhibited in circumstances in which the cortical rotation occurs but certain cytoplasmic components are damaged (Elinson and Pasceri, 1989).

B. The Vegetal Microtubule Array

At the time of the cortical rotation, an impressive array of aligned microtubules appears beneath the vegetal surface of the egg (Elinson and Rowning, 1988; Fig. 3A). This is coincident with the main period of microtubule polymerization in the first cell cycle (Elinson, 1985; Fig. 2). The microtubules lie in the direction of rotation (Elinson and Rowning, 1988; Zisckind and Elinson, 1990) and, given the results of inhibitor studies (Manes et al., 1978; Scharf and Gerhart, 1983; Vincent et al., 1987), they are good candidates for a role in the rotation process. The mechanism by which they become oriented may thus be important in establishing the direction of cortical rotation. The processes of microtubule orientation and cortical rotation are probably mutually reinforcing, with aligned microtubules contributing to directed movement and the movement helping to align the microtubules (see Gerhart et al., 1989). We have recently demonstrated a direct link between the microtubules of the vegetal array with those of the enlarging sperm aster by staining microtubules on egg sections (Houliston and Elinson, 1991a). Astral microtubules extend from the animal cytoplasm into the vegetal cytoplasm of the egg, and then turn to run parallel to the vegetal cortex (Fig. 3B and C). The region of the yolk-free vegetal cortex of the egg favors microtubule polymerization in comparison to the animal cortex, and the eccentric position of the sperm aster introduces asymmetry to the pattern of microtubule polymerization. Once the cortex has started to move relative to the cytoplasm, microtubule

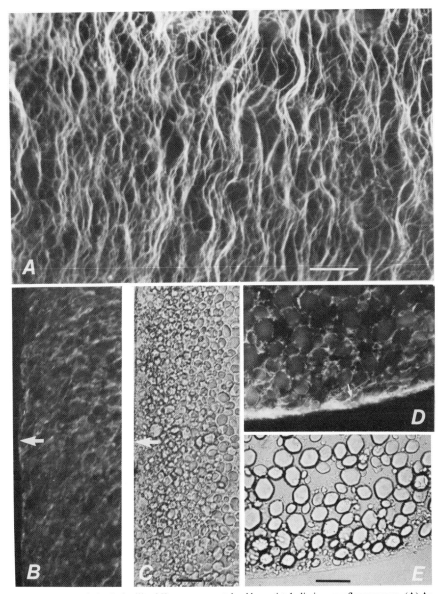

Fig. 3 Microtubules in fertilized *Xenopus* eggs stained by anti-tubulin immunofluorescence. (A) An array of aligned microtubules on a piece of vegetal cortex isolated from an egg half-way through the first cell cycle. (B–E) Sections taken in the plane of the animal and vegetal poles and sperm entry site. Eggs were fixed at the time the array was starting to form (B) or when it was fully formed (D). (C) and (E) are the corresponding bright field images. The area shown in (B) and (C) is at the level of the animal–vegetal cortical pigment boundary (arrowed) on the sperm entry side of the egg. Note the extension of sperm aster microtubules into the vegetal cytoplasm, and the way they turn to run parallel to the vegetal cortex. The section in (D) and (E) is in the same plane but at the vegetal pole. Note the connections between cytoplasmic and cortical microtubules (e.g., at arrow). Scale bars, 25 μm.

ends arriving at the vegetal cell surface should tend to bend in the direction of movement, thus reinforcing the alignment of the array.

The orientation of the dorsal–ventral axis (Ancel and Vintemberger, 1948; Black and Gerhart, 1985) and of the microtubule array (Zisckind and Elinson, 1990) can be aligned by centrifuging or tilting the egg with respect to gravity prior to the time of cortical rotation, overriding the cue from the sperm aster. These treatments may either introduce cytoplasmic asymmetries that affect the growth of the radial microtubules or start movement of the cytoplasm relative to cortex and thus align the microtubules as they arrive in the cortical region. Both cortical rotation (Vincent et al., 1986) and the formation of the vegetal cortical microtubule array (Elinson and Rowning, 1988) can occur in the absence of a sperm aster. Microtubules in such eggs grow outward from the animal cytoplasm (see Section II,B), and random differences in the cytoplasmic organization are presumably able to introduce sufficient asymmetry in their growth to mimic the situation in fertilized eggs.

Certain treatments that cause premature polymerization of the bulk of cytoplasmic microtubules cause a reversal of the direction of cortical rotation. Raising the temperature of eggs briefly early in the cell cycle is particularly effective (Sakai, 1990), having a more uniform effect on both the cortical rotation and axial development than treatment with heavy water (Scharf et al., 1989). It would be interesting to know how this treatment affects the formation of the vegetal microtubule array.

C. Possible Mechanisms of Force Generation

How could the array of aligned microtubules in the vegetal cortex be involved in the generation of movement? Some possibilities are shown in Fig. 4. The occurrence of directed microtubule polymerization, outward from cortex to cytoplasm, raises the possibility that some of the force of the cortical rotation could be provided by the polymerization itself (Fig. 4A). Rapid microtubule growth in the subcortical region might create force between the cortex and cytoplasm, with the oblique angle of approach of the microtubules to the vegetal cortex directing the force (see Fig. 3B and C). Such a phenomenon has been observed in sea urchin eggs. A heat-induced rotation of cytoplasm relative to the cortex is accompanied by polymerization first of a large microtubule aster and then of a spirally arranged cortical array (Harris et al., 1980; Schroeder and Battaglia, 1985). In the frog egg, the force of microtubule polymerization could provide sustained force only if localized in some way, because there is net depolymerization of egg microtubules during the second half of the cell cycle (Elinson, 1985).

Another proposal for force generation is that the microtubules of the array may act as tracks along which translocation occurs mediated by a microtubule-associated motor molecule (Vincent et al., 1987; Elinson and Rowning, 1988; Fig.

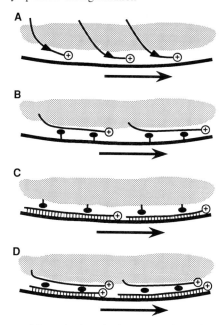

Fig. 4 Diagram showing possible mechanisms for force generation during the cortical rotation. In each case the shear occurs between the cytoplasmic mass (stippled) and the vegetal cortex (thick line). Microtubules are oriented predominantly with their plus ends running in the direction of cortical rotation, indicated by the large arrow below each drawing. (A) Force could be generated by directional polymerization (small arrows) of microtubules into the cortical region. (B–D) Microtubule-associated motor molecules (black blobs) could generate movement between microtubules and cortex (B), between microtubules and cytoplasm (C), or between different populations of microtubules (D), depending on where the microtubules are anchored.

4B–D). Because the direction of movement created by such molecules depends on the polarity of microtubules, we were motivated to determine the polarity of these microtubules using the "hook assay" (Heidemann and McIntosh, 1980). We found that the vast majority of the microtubules run in the same direction. The hooks indicated that about 90% of cortical microtubules have their plus ends facing the gray crescent, that is, pointing in the direction of cortical movement (Houliston and Elinson, 1991b; Fig. 5). This situation is due in part to the extension of microtubule plus ends out through the cytoplasm during the formation of the array and their turning in the direction of rotation at the vegetal cortex (Houliston and Elinson, 1991a; Fig. 3B and C).

The implication of this polarity for the mechanism of force generation depends on how the microtubules of the array are anchored at the time of rotation. If they are anchored to the cytoplasmic mass (Fig. 4B), a plus end-directed motor such as kinesin could carry the cortex in the right direction. Immunofluorescence performed with monoclonal antibodies to sea urchin kinesin suggests that there is

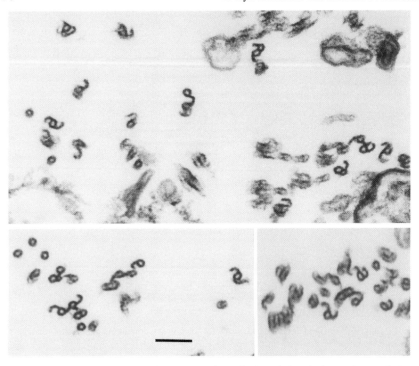

Fig. 5 Electron micrographs showing cross-sections of microtubules of pieces of cortex isolated from *Rana pipiens* eggs at the time of the cortical rotation. "Hooks" on the microtubules created by the addition of exogenous tubulin indicate the polarity of the microtubules, with anticlockwise hooks representing microtubules with their plus ends pointing away from the viewer. In this cortex, the view is toward the gray crescent (three different regions). Scale bar, 0.1 μm.

a frog egg kinesin associated with both the aligned microtubules and endoplasmic reticulum in the cortical region (Houliston and Elinson, 1991b), so we can suggest that kinesin attached to ER on the cortex creates the movement. If, however, the microtubules are anchored to the cortex (Fig. 4C), minus end-directed microtubule motor molecules might create movement, provided they are somehow attached to the cytoplasmic mass. This possibility seems less likely on current evidence because some cortical microtubules are found to be continuous with ones in the cytoplasm throughout the period of rotation (Houliston and Elinson, 1991a; Fig. 3D and E). A third possibility is that microtubule motors, perhaps related to dynein or dynamin (see Vallee and Shpetner, 1990), promote sliding between different populations of microtubules, some anchored to the cortex and others to the cytoplasm (Fig. 4D).

One way to distinguish these possibilities experimentally might be to exploit the fact that pieces of vegetal cortex with aligned microtubules attached can be isolated cleanly from the egg (Houliston and Elinson, 1991b; Fig. 3A). Judicious application of different nucleotides, antibodies, and inhibitors may allow the cortical translocation to be reconstructed *in vitro* and its mechanism elucidated.

V. Conclusions

Microtubules in the frog egg play essential roles in both cell cycle events (mitosis, cleavage) and key processes of cytoplasmic organization, including the establishment of animal–vegetal polarity and the cortical rotation that specifies the dorsal–ventral axis. Our current understanding of their involvement in these events, and of the mechanisms by which their distribution is regulated, is still quite poor. The matter is worth pursuing because the larger question of the nature of the cytoplasmic information that specifies embryonic body axes is almost certainly tied in with the cell biology of the egg.

Acknowledgments

We thank Paul Clute for useful comments on Section II. Our research is supported by grants to R.P.E. from NSERC and MRC, Canada. E.H. acknowledges the receipt of an SERC/NATO postdoctoral fellowship.

References

Allen, V. J., and Vale, R. D. (1991). Cell cycle control of microtubule-based membrane transport and tubule formation *in vitro*. *J. Cell Biol.* **113**, 347–360.
Al-Mukhtar, K. A. K., and Webb, A. C. (1971). An ultrastructural study of primordial germ cells, oogonia and early oocytes in *Xenopus laevis*. *J. Embryol. Exp. Morphol.* **26**, 195–217.
Ancel, P., and Vintemberger, P. (1948). Recherches sur le déterminisme de la symmétrie bilatérale dans l'oeuf des amphibiens. *Bull. Biol. Fr. Belg., Suppl.* **31**, 1–182.
Belmont, L. D., Hyman, A. A., Sawin, K. E., and Mitchison, T. J. (1990). Real-time visualization of cell cycle-dependent changes in microtubule dynamics in cytoplasmic extracts. *Cell (Cambridge, Mass.)* **62**, 579–589.
Bjerknes, M. (1986). Physical theory of the orientation of astral mitotic spindles. *Science* **234**, 1413–1416.
Black, S. D., and Gerhart, J. C. (1985). Experimental control of the site of axis formation in *Xenopus laevis* eggs centrifuged before first cleavage. *Dev. Biol.* **108**, 310–324.
Briedis, A., and Elinson, R. P. (1982). Suppression of male pronuclear movement in frog eggs by hydrostatic pressure and deuterium oxide yields androgenetic haploids. *J. Exp. Zool.* **222**, 45–57.
Brinkley, B. R., Fuller, G., and Highfield, D. (1976). Tubulin antibodies as probes for microtubules in dividing and non-dividing mammalian cells. *In* "Cell Motility" (R. Goldman, T. Pollard, and J. Rosenbaum, eds.), pp. 435–456. Cold Spring Harbor Lab, Cold Spring Harbor, New York.
Busa, W. B., and Nuccitelli, R. (1985). An elevated free cytosolic Ca^{2+} wave follows fertilization in eggs of the frog, *Xenopus laevis*. *J. Cell Biol.* **100**, 1325–1329.

Byers, T. J., and Armstrong, P. B. (1986). Membrane protein redistribution during *Xenopus* first cleavage. *J. Cell Biol.* **102,** 2176–2184.

Clavert, J. (1962). Symmetrization of the egg of vertebrates. *Adv. Morphol.* **2,** 27–60.

Clute, P. (1989). The effect of microtubule inhibitors on chromosome cycles of *Xenopus laevis* blastomeres. MSc. Thesis, University of Toronto.

Colman, A., Morser, J., Lane, C., Besley, J., Wylie, C. C., and Valle, G. (1981). Fate of secretory proteins trapped in oocytes of *Xenopus laevis* by disruption of the cytoskeleton or by imbalanced subunit synthesis. *J. Cell Biol.* **91,** 770–780.

Dent, J. A., and Klymkowsky, M. W. (1989). Whole mount analyses of cytoskeletal reorganization and function during oogenesis and embryogenesis in *Xenopus*. In "The Cell Biology of Fertilization" (H. Schatten and G. Schatten, eds.), pp. 63–103. Academic Press, San Diego, California.

Dumont, J. N., and Wallace, R. A. (1972). The effects of vinblastine on isolated *Xenopus* oocytes. *J. Cell Biol.* **53,** 605–610.

Elinson, R. P. (1980). The amphibian egg cortex in fertilization and early development. *Symp. Soc. Dev. Biol.* **38,** 217–234.

Elinson, R. P. (1983). Cytoplasmic phases in the first cell cycle of the activated frog egg. *Dev. Biol.* **100,** 440–451.

Elinson, R. P. (1985). Changes in levels of polymeric tubulin associated with activation and dorsoventral polarization of the frog egg. *Dev. Biol.* **109,** 224–233.

Elinson, R. P. (1990). Cortical and cytoplasmic phases in amphibian eggs. *Ann. N. Y. Acad. Sci.* **582,** 31–39.

Elinson, R. P., and Houliston, E. (1990). Cytoskeleton in *Xenopus* oocytes and eggs. *Semin. Cell Biol.* **1,** 349–357.

Elinson, R. P., and Kao, K. R. (1989). The location of dorsal information in frog early development. *Dev. Growth Differ.* **31,** 423–430.

Elinson, R. P., and Pasceri, P. (1989). Two UV-sensitive targets in dorsoanterior specification of frog embryos. *Development (Cambridge, UK)* **106,** 511–518.

Elinson, R. P., and Rowning, B. (1988). A transient array of parallel microtubules in frog eggs: Potential tracks for a cytoplasmic rotation that specifies the dorso-ventral axis. *Dev. Biol.* **128,** 185–197.

Gard, D. L. (1991). Organization, nucleation, and acetylation of microtubules in *Xenopus laevis* oocytes: A study by confocal immunofluorescence microscopy. *Dev. Biol.* **143,** 346–362.

Gard, D. L., and Kirschner, M. (1987a). Microtubule assembly in cytoplasmic extracts of *Xenopus* oocytes and eggs. *J. Cell Biol.* **105,** 2191–2201.

Gard, D. L., and Kirschner, M. (1987b). A microtubule-associated protein from *Xenopus* eggs that specifically promotes assembly at the plus-end. *J. Cell Biol.* **105,** 2203–2215.

Gard, D. L., Hafezi, S., Zhang, T., and Doxsey, S. J. (1990). Centrosome duplication continues in cycloheximide-treated *Xenopus* blastulae in the absence of a detectable cell cycle. *J. Cell Biol.* **110,** 2033–2042.

Gerhart, J. C., Wu, M., and Kirschner, M. (1984). Cell-cycle dynamics of an M-phase-specific cytoplasmic factor in *Xenopus laevis* oocytes and eggs. *J. Cell Biol.* **98,** 1247–1255.

Gerhart, J. C., Danilchik, M., Doniach, T., Roberts, S., Rowning, B., and Stewart, R. (1989). Cortical rotation of the *Xenopus* egg: Consequences for the anteroposterior pattern of embryonic dorsal development. *Development (Cambridge, UK)* **107,** Suppl., 37–51.

Hara, K., Tydeman, P., and Kirschner, M. (1980). A cytoplasmic clock with the same period as the division cycle in *Xenopus* eggs. *Proc. Natl. Acad. Sci. U.S.A.* **77,** 462–466.

Harris, P., Osborne, M, and Weber, K. (1980). A spiral array of microtubules in the fertilized sea urchin egg cortex examined by indirect immunofluorescence and electron microscopy. *Exp. Cell Res.* **126,** 227–236.

Heasman, J., Quarmby, J., and Wylie, C. C. (1984). The mitochondrial cloud of *Xenopus* oocytes: The source of germinal granule material. *Dev. Biol.* **105,** 458–469.

Heidemann, S. R., and Kirschner, M. W. (1978). Induced formation of asters and cleavage furrows in oocytes of *Xenopus laevis* during *in vitro* maturation. *J. Exp. Zool.* **204**, 431–444.

Heidemann, S. R., and McIntosh, J. R. (1980). Visualization of the structural polarity of microtubules, *Nature (London)* **286**, 517–519.

Houliston, E., and Elinson, R. P. (1991a). Patterns of microtubule polymerization relating to cortical rotation in *Xenopus laevis* eggs. *Development (Cambridge, UK)* **112**, 107–117.

Houliston, E., and Elinson, R. P. (1991b). Evidence for the involvement of microtubules, ER, and kinesin in the cortical rotation of fertilized frog eggs. *J. Cell Biol.*, **114**, 1017–1028.

Huchon, D., Jessus, C., Thibier, C., and Ozon, R. (1988). Presence of microtubules in isolated cortices of prophase I and metaphase II oocytes in *Xenopus laevis*. *Cell Tissue Res.* **254**, 415–442.

Jessus, C., Thibier, C., and Ozon, R. (1985). Identification of microtubule-associated proteins (MAPs) in *Xenopus* oocyte. *FEBS Lett.* **192**, 135–140.

Jessus, C., Huchon, D., and Ozon, R. (1986). Distribution of microtubules during the breakdown of the nuclear envelope of the *Xenopus* oocyte: An immunocytochemical study. *Biol. Cell.* **56**, 113–120.

Jessus, C., Thibier, C., and Ozon, R. (1987). Levels of microtubules during the meiotic maturation of the *Xenopus* oocyte. *J. Cell Sci.* **87**, 705–712.

Karsenti, E., Newport, J., Hubble, R., and Kirschner, M. (1984). Interconversion of metaphase and interphase microtubule arrays as studied by the injection of centrosomes and nuclei into *Xenopus* eggs. *J. Cell Biol.* **98**, 1730–1745.

Keller, R. E. (1986). The cellular basis of amphibian gastrulation. *In* "Developmental Biology: A Comprehensive Synthesis" (L. W. Browder, ed.), Vol. 2, pp. 241–384. Plenum, New York.

Kirschner, M., and Mitchison, T. (1986). Beyond self-assembly: From microtubules to morphogenesis. *Cell (Cambridge, Mass.)* **45**, 329–342.

Klymkowsky, M. W., Maynell, L. A., and Polson, A. G. (1987). Polar asymmetry in the organization of the cortical cytokeratin system of *Xenopus laevis* oocytes and embryos. *Development (Cambridge, UK)* **100**, 543–557.

Kobayakawa, Y. (1988). Role of mitotic asters in accumulation of pigment granules around nuclei in early amphibian embryos. *J. Exp. Zool.* **248**, 232–237.

Kubota, H. Y., Yoshimoto, Y., Yoneda, M., and Hiramoto, Y. (1987). Free calcium wave upon activation in *Xenopus* eggs. *Dev. Biol.* **119**, 129–136.

Lessman, C. A. (1987). Germinal vesicle migration and dissolution in *Rana pipiens* oocytes: Effects of steroids and microtubule poisons. *Cell Differ.* **20**, 238–251.

Malacinski, G. M., Benford, H., and Chung, H. M. (1975). Association of an ultra-violet irradiation sensitive cytoplasmic localization with the future dorsal side of the amphibian egg. *J. Exp. Zool.* **191**, 97–110.

Manes, M. E., and Barbieri, F. D. (1977). On the possibility of sperm aster involvement in dorso-ventral polarization and pronuclear migration in the amphibian egg. *J. Embryol. Exp. Morphol.* **40**, 187–197.

Manes, M. E., and Elinson, R. P. (1980). Ultra-violet light inhibits grey crescent formation on the frog egg. *Wilhelm Roux's Arch. Dev. Biol.* **189**, 73–76.

Manes, M. E., Elinson, R. P., and Barbieri, F. D. (1978). Formation of the amphibian grey crescent: Effects of colchicine and cytochalasin B. *Wilhelm Roux's Arch. Dev. Biol.* **185**, 99–104.

Maro, B., Howlett, S. K., and Houliston, E. (1986). Cytoskeletal dynamics in the mouse egg. *J. Cell Sci. Suppl.* **5**, 343–359.

Melton, D. A. (1987). Translocation of a localized maternal mRNA to the vegetal pole of *Xenopus* oocytes. *Nature (London)* **328**, 80–82.

Murray, A., and Kirschner, M. (1989). Cyclin synthesis drives the early embryonic cell cycle. *Nature (London)* **339**, 275–280.

Neighbors, B. W., Williams, R. C., and McIntosh, J. R. (1988). Localization of kinesin in cultured cells. *J. Cell Biol.* **106**, 1193–1204.

Paleček, J., Habrová, V., Nedvídek, J., and Romanovsky, A. (1985). Dynamics of tubulin structures in *Xenopus laevis* oogenesis. *J. Embryol. Exp. Morphol.* **87,** 75–86.

Ressom, R. E., and Dixon, K. E. (1988). Relocation and reorganization of germ plasm in *Xenopus* embryos after fertilization. *Development (Cambridge, UK)* **103,** 507–518.

Sakai, M. (1990). Heat-induced reversal of dorsal-ventral polarity in *Xenopus* eggs. *Dev., Growth Differ.* **32,** 497–504.

Satoh, H., and Shinagawa, A. (1990). Mechanism of dorso-ventral axis specification in nuclear transplanted eggs of *Xenopus laevis. Dev., Growth Diff.* **32,** 609–617.

Scharf, S. R., and Gerhart, J. C. (1980). Determination of the dorso-ventral axis in eggs of *Xenopus laevis:* Complete rescue of uv-impaired eggs by oblique orientation before first cleavage. *Dev. Biol.* **79,** 181–198.

Scharf, S. R., and Gerhart, J. C. (1983). Axis determination in eggs of *Xenopus laevis:* A critical period before first cleavage identified by the common effects of cold, pressure, and uv-irradiation. *Dev. Biol.* **99,** 75–87.

Scharf, S. R., Rowning, B., and Gerhart, J. C. (1989). Hyperdorsoanterior embryos from *Xenopus* eggs treated with D$_2$O. *Dev. Biol.* **134,** 175–188.

Schroeder, T. E., and Battaglia, D. E. (1985). "Spiral asters" and cytoplasmic rotation in sea urchin eggs: Induction in *Strongylocentrotus purpuratus* eggs by elevated temperature. *J. Cell Biol.* **100,** 1056–1062.

Sluder, G. (1979). Role of spindle microtubules in the control of cell cycle timing. *J. Cell Biol.* **80,** 674–691.

Stewart-Savage, J., and Grey, R. D. (1982). The temporal and spatial relationships between cortical contraction, sperm trail formation, and pronuclear migration in fertilized *Xenopus* eggs. *Wilhelm Roux's Arch. Dev. Biol.* **191,** 241–245.

Ubbels, G. A., Hara, K., Koster, C. H., and Kirschner, M. W. (1983). Evidence for a functional role for the cytoskeleton in determination of the dorsoventral axis in *Xenopus laevis. J. Embryol. Exp. Morphol.* **77,** 15–37.

Vale, R. D. (1991). Mitotically activated microtubule severing protein in *Xenopus* egg extracts. *Cell* **64,** 827–839.

Vallee, R. B., and Bloom, G. S. (1991). Mechanisms of fast and slow axonal transport. *Annu. Rev. Neurosci.* **14,** 59–92.

Vallee, R. B., and Shpetner, H. S. (1990). Motor proteins of cytoplasmic microtubules. *Annu. Rev. Biochem.* **59,** 909–932.

Verde, F., Labbé, J.-C., Dorée, M., and Karsenti, E. (1990). Regulation of microtubule dynamics by cdc2 protein kinase in cell-free extracts of *Xenopus* eggs. *Nature (London)* **343,** 233–238.

Vincent, J.-P., Oster, G. F., and Gerhart, J. C. (1986). Kinematics of gray crescent formation in *Xenopus* eggs. *Dev. Biol.* **113,** 484–500.

Vincent, J.-P., Scharf, S. R., and Gerhart, J. C. (1987). Subcortical rotation in *Xenopus* eggs: A preliminary study of its mechanochemical basis. *Cell Motil. Cytoskel.* **8,** 143–154.

Wasserman, W., and Smith, L. D. (1978). The cyclic behaviour of a cytoplasmic factor controlling nuclear envelope breakdown. *J. Cell Biol.* **78,** R15–R22.

Wylie, C. C., Brown, D., Godsave, S. F., and Quarmby, J. (1985). The cytoskeleton of *Xenopus* oocytes and its role in development. *J. Embryol. Exp. Morphol.* **89,** Suppl. 1–15.

Yisraeli, J. K., Sokol, S., and Melton, D. A. (1990). A two-step model for the localization of maternal mRNA in *Xenopus* oocytes: Involvement of microtubules and microfilaments in the translocation and anchoring of Vg1 mRNA. *Development (Cambridge, UK)* **108,** 289–298.

Zisckind, N., and Elinson, R. P. (1990). Gravity and microtubules in dorsoventral polarization of the *Xenopus* egg. *Dev., Growth Differ.* **32,** 575–581.

5

Microtubule Motors in the Early Sea Urchin Embryo

Brent D. Wright and Jonathan M. Scholey
Department of Zoology
University of California at Davis
Davis, California 95616
and
Department of Cellular and Structural Biology
University of Colorado Health Science Center
Denver, Colorado 80262

I. Summary

Sea urchin gametes and early embryos have proven to be a useful system for studying the roles of microtubule (MT)-associated motors in axonemal motility and cytoplasmic MT-based movements in dividing cells. In this brief article, known and potential sea urchin MT motors are listed and their possible biological functions are discussed.

II. Sea Urchin Embryos, Microtubule Motility, and Motors

A. Microtubule Motility

Microtubule-associated motor proteins drive axonemal motility and the intracellular transport of various cargos, including membranous organelles and chromosomes, by coupling the hydrolysis of ATP to the generation of force and

72   Brent D. Wright and Jonathan M. Scholey

motion relative to microtubules (MTs). Axonemal dyneins are MT-motor proteins that drive flagellar and ciliary motility, while three classes of cytoplasmic motors, namely cytoplasmic dyneins, kinesins and dynamin, are thought to drive motions relative to cytoplasmic MTs (for reviews, see McIntosh and Porter, 1989; Vallee and Shpetner, 1990).

The arrays of MTs that serve as tracks along which the MT motors move usually radiate from MT-organizing centers (MTOCs), such as basal bodies, centrosomes, and spindle poles, with the fast-polymerizing plus end of each polar MT pointing away from the MTOC (Gibbons, 1981; McIntosh, 1983; Kirschner and Mitchison, 1986). Therefore, motor proteins that track toward the plus ends of MTs usually move away from the MTOC, whereas those moving toward the MTOC display minus end-directed transport (Vale, 1987, 1990; McIntosh and Porter, 1989).

Video microscopic motility assays have been extremely useful in characterizing the functional properties of motor proteins *in vitro*. Both axonemal (Gibbons, 1965) and cytoplasmic (Lye *et al.*, 1987; Paschal and Vallee, 1987) forms of the MT-motor dynein generally move objects along MTs toward their minus ends *in vitro* (Fox and Sale, 1987; Paschal *et al.*, 1987; Paschal and Vallee, 1987), although in one case (Euteneuer *et al.*, 1988) cytoplasmic dynein apparently mediates bidirectional transport. Axonemal dyneins drive sliding of outer doublet MTs relative to one another (Gibbons, 1981) while cytoplasmic dyneins may participate in vesicle transport (Schroer *et al.*, 1989; Schnapp and Reese, 1989) and kinetochore-based chromosome movements (Pfarr *et al.*, 1990; Steuer *et al.*, 1990).

The MT-motor dynamin drives active sliding between microtubules *in vitro* (Shpetner and Vallee, 1989) and may cause MT–MT sliding in the cytoplasm during interphase (Koonce *et al.*, 1987) and in the zone of MT overlap within the mitotic spindle (Masuda *et al.*, 1988).

A third MT motor, kinesin (Vale *et al.*, 1985a), transports particles toward the plus ends of MTs *in vitro* (Vale *et al.*, 1985b; Porter *et al.*, 1987) and appears to move membranous vesicles and organelles along MTs in cells (Schroer *et al.*, 1988; Brady *et al.*, 1990; Hollenbeck and Swanson, 1990; Saxton *et al.*, 1991; Wright *et al.*, 1991). Recent genetic studies have identified several related kinesin-like proteins (Endow *et al.*, 1990; McDonald and Goldstein, 1990; Zhang *et al.*, 1990; Enos and Morris, 1990; Meluh and Rose, 1990; Hagan and

Fig. 1 Time sequence showing normal development of *Strongylocentrotus purpuratus* embryos at 15°C. (a) Fertilized egg with elevated fertilization envelope. The egg diameter is 80 μm. (b) The 16-cell embryo (5.5 hr after fertilization). (c) Mesenchyme blastula (18 hr after fertilization). Hatching occurs at about this stage. (d) Late gastrula (46 hr after fertilization). The embryos are covered with cilia and are free swimming. (e) Early pluteus (72 hr after fertilization), ventral view. (f) Fed late pluteus larva (3 weeks after fertilization). Approximate magnifications: (a to c) × 375; (d to f) progressively decreasing to × 75. Adapted from Davidson *et al.* (1982; Fig. 1); copyright 1982 by the AAAS.

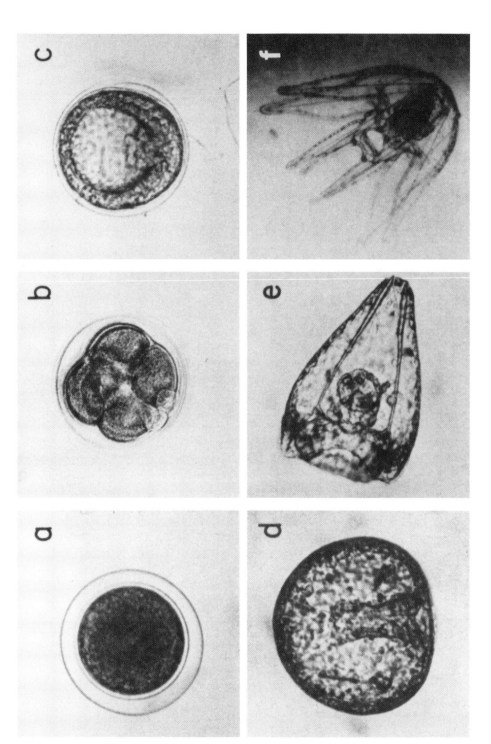

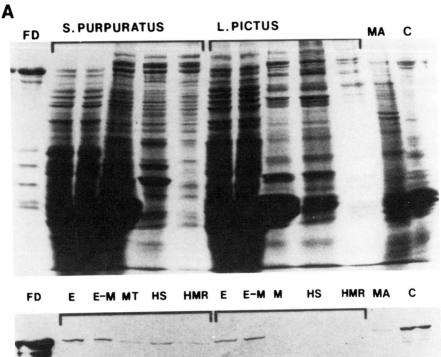

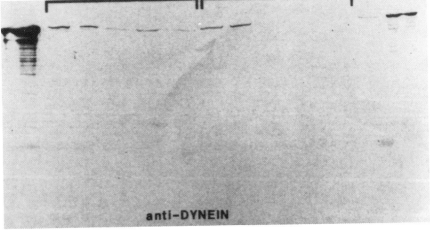

B

Fig. 2   Distribution of dynein-related polypeptides in different sea urchin egg fractions. The following fractions were run on a 5–15% polyacrylamide gradient gel (A), blotted to nitrocellulose, and probed with an affinity-purified antibody to flagellar dynein (B). The lanes from left to right were as follow: FD, flagellar dynein; from both *S. purpuratus* and *Lytechinus pictus* eggs: E, egg extract; E-M, microtubule-depleted supernatant; MT (or M), microtubule pellet; HS, high-salt extract of MTs; HMR, high-molecular-weight MAPs; then MA, isolated mitotic apparatus of *S. purpuratus;* C, embryonic cilia of *S. purpuratus*. (From Porter *et al.* 1988; Fig. 9.)

Yanagida, 1990; reviewed in Vale and Goldstein, 1990) that appear to be MT motors important for meiotic and mitotic chromosome movements. These kinesin-like motors display sequence similarity to "conventional" kinesins within their motor domains, although in at least one case (*Drosophila ncd* or *claret*) the kinesin-like motor drives minus end-directed MT motility *in vitro* (Walker *et al.*, 1990; McDonald *et al.*, 1990), as does dynein.

B. Sea Urchin Microtubule Motors

Echinoderm gametes and embryos have proven to be a useful system for dissecting the roles of MT motors, particularly in axonemal motility and cell division (Gibbons, 1975; Bloom and Vallee, 1989; Scholey *et al.*, 1985; Porter *et al.*, 1988). Sea urchins routinely yield large quantities of gametes that can produce populations of rapidly and synchronously dividing embryos (Schroeder, 1986; Fig. 1), which have stockpiled large quantities of maternally loaded proteins required for the numerous rapid divisions of the cleavage stage (Davidson *et al.*,

Table I Microtubule Motors from Sea Urchin Gametes and Embryos[a]

Motor protein	Description	Function
Axonemal dyneins	Outer arm (21S): 2 HCs (~400 kd); two headed; ATP-sensitive MT-binding; ATPase is MT and TX-100 stimulated and vanadate inhibited; HUV/LUV (~200 kd); (−)end-directed MT-motor	Axonemal beating via outer MT-doublet sliding
	Inner arm: five unique HCs; two forms?; HUV/LUV and ATPase like outer arm; (−)end-directed MT-motor	Similar to outer arm
Egg cytoplasmic dyneins (20S)	>90% = weak MT-binding; cross-reacts with flagellar dynein antibodies; HCs, HUV/LUV and ATPase like axonemal dyneins	Inactive ciliary dynein precursor?
	<10% = ATP-sensitive MT-binding; weak cross-reactivity with flagellar dynein; unique HC(s) (HM$_r$3) and HUV/LUV; low specific ATPase activity	Inward-directed movements along cytoplasmic MTs?
Kinesin (9.6S)	Nucleotide-sensitive MT-binding; HCs:LCs (130:84/78 kDa); unique MT-ATP/GTPase; two-headed; (+)end-directed MT-translocator with substrates and inhibitors of motility well characterized	Outward-directed movements along cytoplasmic MTs?
10S MT-ATPase	ATP-insensitive MT-binding; unique MT-ATP/GTPase	Dynamin-like, mediating MT–MT sliding?

[a]HC, Heavy chain; LC, light chain; HUV/LUV, heavy and light UV/vanadate cleavage products; MT, microtubule; MT-ATP/GTPase, MT-stimulated ATP/GTPase; TX-100, Triton X-100.

1982). This allows the isolation of sperm flagellar or embryonic ciliary axonemes, cytoplasmic MTs, and stage-specific mitotic apparatuses (MAs) in sufficient quantities for biochemical analysis of their MT-associated proteins (MAPs)
(see Fig. 2 for example, and Porter *et al.*, 1988). Thus it is possible to isolate
biochemical quantities of MT-associated motor proteins that may play important
roles in dividing cells or in flagellar and ciliary axonemes.

    Furthermore, it is possible to reconstitute axonemal motility in extracted ax-
onemes (Gibbons and Gibbons, 1972; reviewed in Porter and Johnson, 1989),
and to study particle transport and MT gliding in cell extracts (Pryer *et al.*, 1986)
and in transport systems reconstituted from isolated components (Cohn *et al.*,
1989). In addition, the clarity of these embryos facilitates the immunocytochemical localization of MT-motor proteins to specific intracellular components (e.g., Scholey *et al.*, 1985; Wright *et al.*, 1991). Finally, the ability to
microinject (Kiehart, 1982) function-blocking antibodies into dividing
blastomeres (Mabuchi and Okuno, 1977; Kiehart *et al.*, 1982; Hafner and Petzelt, 1987; Dinsmore and Sloboda, 1989) provides a potential means of directly
identifying the biological roles of motor proteins in intracellular transport.

    Using this system, flagellar and ciliary dyneins (Table I) have been extensively
studied and their roles in axonemal motility thoroughly characterized (reviewed
in Porter and Johnson, 1989). In addition, several demonstrated and candidate
MT motors have been isolated from egg cytosol, including cytoplasmic dyneins,
kinesin, and a 10S MT-activated ATPase (Table I), which may prove to have
important functions in mitotic cells.

III. Dyneins: Biochemistry and Motility

A. Axonemal Dyneins

Dyneins were first identified as the inner and outer arm proteins of MTs in sea
urchin flagellar and ciliary axonemes (Gibbons, 1965, 1981; Ogawa *et al.*, 1977;
Kimura, 1977; see Fig. 2) and were shown to cause flagellar beating by generating shear forces between adjacent outer doublet MTs (Gibbons and Gibbons,
1972, 1973, 1976, 1979; Summers and Gibbons, 1973; B. H. Gibbons *et al.*,
1976; I. R. Gibbons *et al.*, 1976; Fox and Sale, 1987; Paschal *et al.*, 1987; Vale
et al., 1989; for detailed reviews, see Porter and Johnson, 1989; Warner *et al.*,
1989; Fig. 3).

    These dyneins can be distinguished from other MT motors by the polarity of
their motility, their biochemistry, and pharmacology (see Table I). The large size
of the native dynein particle (primarily 21S) and heavy chains (HCs) (about 400
kDa) is distinctive. Their ATPase activity is MT stimulated, can be enhanced by
nonionic detergents, and is potently inhibited by vanadate (<10 μM). Near-UV
irradiation (365 nm) in the presence of $Mg^{2+} \cdot ATP$ and vanadate specifically

cleaves the dynein HC into two fragments (HUV and LUV; about 200 kDa each), resulting in loss of motility and ATPase activity (B. H. Gibbons and Gibbons, 1987; I. R. Gibbons and Gibbons, 1987). Different dynein isoforms can be discriminated from each other by the slightly different sizes of their HCs and the resultant UV/vanadate cleavage products, the lack of antibody cross-reactivity between their HCs, and differences in their enzyme properties and pharmacologic sensitivities.

Generally, sea urchin flagellar and the less well-characterized ciliary dyneins are similar (Kimura, 1977; see Fig. 2), but their individual outer and inner arm dyneins appear to be heterologous both structurally and in terms of their polypeptide composition (I. R. Gibbons et al., 1976; Ogawa et al., 1982; for reviews, see Gibbons, 1988; Porter and Johnson, 1989). Sea urchin flagellar outer arm dyneins are two headed and have two distinct HCs (α and β; I. R. Gibbons and Gibbons, 1987; Sale et al., 1985), while the more complex and less well-understood inner arm dyneins apparently have two forms and as many as five different HCs (Sale and Fox, 1986; Sale et al., 1989). Furthermore, monoclonal antibodies to outer arm HCs, which recognize homologous outer arm HCs, even across species, do not cross-react with the heterologous outer arm HC or with any of the inner arm HCs from the same species (Piperno, 1984; Mocz et al., 1988).

However, outer and inner arm dyneins appear to be functionally very similar, with each HC corresponding to a single head that acts as a minus end-directed MT motor in situ and in vitro (Gibbons and Gibbons, 1973, 1976; Fox and Sale, 1987; Paschal et al., 1987; Vale et al., 1989; Fig. 3). Each of these motors utilizes the energy derived from ATP hydrolysis, via cyclic cross-bridge formation and conformational changes, to generate sliding forces between the A-subfiber of one outer MT doublet to which it is attached and the B-tubule of the adjacent doublet within the axoneme (Summers and Gibbons, 1973; Fox and Sale, 1987).

B. Cytoplasmic Dyneins

A dynein-like ATPase activity was first identified in sea urchin eggs by Weisenberg and Taylor (1968). Several subsequent studies revealed the presence of multiple isoforms (reviewed by Lye et al., 1989; Pratt, 1989) that have been resolved into two classes of cytosolic dyneins (Porter et al., 1988; see Fig. 2 and Table I). Both particles sediment primarily at 20S on sucrose gradients but are separated by MT-affinity binding. The most abundant isoform has a low affinity for MTs and thus may be biologically inactive, but its antibody reactivity, polypeptide composition, and enzymatic activity (listed in Table I) indicate that it is closely related to axonemal dynein. This protein is thought to represent a maternally loaded cytoplasmic precursor of ciliary dynein (Asai, 1986; Porter et al.,

Brent D. Wright and Jonathan M. Scholey

A

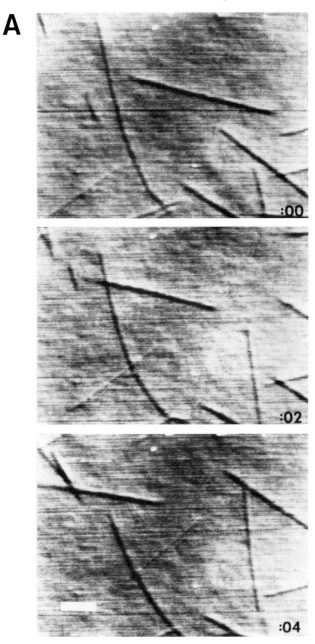

Fig. 3

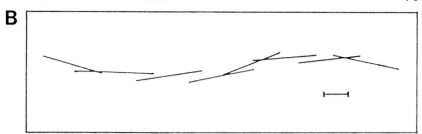

Fig. 3 Dynein-mediated microtubule gliding. *Strongylocentrotus purpuratus* sea urchin sperm outer-arm dynein purified by sucrose gradient centrifugation was applied to a coverslip and examined for motility. (A) Eight microtubules were observed to glide within this field in the 4-sec interval shown. Scale bar, 1 μm. (B) Video images of a single microtubule were traced over a 12-sec interval to show the lateral mobility during translocation. Scale bar, 1 μm. (Adapted from Paschal *et al.*, 1987, Fig. 2; reprinted by permission from *Nature;* copyright 1987 Macmillan Magazines Limited.)

1988), consistent with the findings of Auclair and Siegel (1966), who showed that urchin embryos possessed sufficient protein stockpiles to reciliate in the absence of protein synthesis.

A different dynein isoform, cosedimented with MTs, was released by ATP or high salt, and possessed distinct HC polypeptide(s) (by high-resolution gels, weak reactivity with flagellar dynein HC antibodies, and its pattern of UV/vanadate cleavage) (designated as HM_r3; Scholey *et al.*, 1984; Porter *et al.*, 1988; see Fig. 2). It is distinct from either axonemal dynein or the ciliary precursor pool, as shown by its unique HC composition and its much lower specific ATPase activity. Although neither sea urchin cytoplasmic dynein isoform has demonstrated MT motility, this less abundant (nonciliary) isoform is a likely candidate for a bona fide cytoplasmic dynein (Lye *et al.*, 1987; Paschal and Vallee, 1987) of dividing cleavage-stage blastomeres.

IV. Kinesin: Biochemistry and Motility

We have isolated and purified cytosolic sea urchin kinesin from eggs and early embryos (Scholey *et al.*, 1985) and have extensively characterized its subunit composition, HC sequence, structure, localization, and MT-motility properties (Porter *et al.*, 1987; Ingold *et al.*, 1988; Cohn *et al.*, 1989; Scholey *et al.*, 1989; Johnson *et al.*, 1990; Wright *et al.*, 1991; Buster and Scholey, 1991). Native sea urchin kinesin is an MT-stimulated ATPase that displays nucleotide-sensitive MT binding, sediments at 9.6S, and is a heterotetramer consisting of two HCs (130 kDa) and two light chains (LCs) (84/78-kDa doublet). Structurally, the HCs are tripartite, forming two globular amino-terminal head domains, which are motor domains containing the ATP- and MT-binding sites, attached via a coiled coil α-helical stalk to a bipartite globular carboxy-terminal tail that, by analogy with

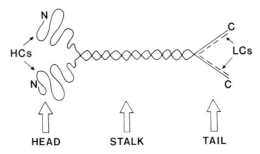

Fig. 4   Diagram of sea urchin kinesin macromolecular structure. Kinesin is a heterotetramer of two heavy chains forming two globular amino-terminal heads, each containing MT and nucleotide-binding sites and functioning as motor domains, connected by a coiled coil α-helical stalk to a bipartite globular tail with two associated light chains, presumably the site of attachment to the transported ligand. (Adapted from a diagram by Karen P. Wedaman, University of California, Davis.)

bovine brain kinesin (Hirokawa *et al.*, 1989), is believed to be the site of LC association and attachment to the transported ligand (Fig. 4). The deduced amino acid sequence of the HC confirms that this is a conventional kinesin, not a kinesin-like protein (Wright *et al.*, 1991).

In vitro, kinesin utilizes $Mg^{2+} \cdot ATP$ or, less effectively, $Mg^{2+} \cdot GTP$ hydrolysis (apparent $K_m = 60$ μM and 2 mM, respectively) to drive plus end-directed movement along MTs (see Fig. 5) with a V_{max} of about 0.6 μm/sec in ATP and 0.4 μm/sec in GTP. In contrast to dyneins, its HC is not cleaved by UV/vanadate. Its motility is inhibited by some monoclonal antibodies to the motor domain (Ingold *et al.*, 1988), as well as by Mg^{2+} chelators, ADP, and ATPγS, and, in a complex fashion, by AMP-PNP and vanadate. This activity is also weakly inhibited by the complex inhibitors AMP-PCP and *N*-ethylmaleimide (NEM) (Cohn *et al.*, 1989).

V. 10S Microtubule-Activated ATPase

Another ATPase from sea urchin eggs was identified (Collins and Vallee, 1986) that is stimulated by MTs (Fig. 6), predicting an MT-motility role. Unlike cytoplasmic dyneins this activity sediments at 10S on sucrose density gradients, is not inhibited by 100 μM vanadate or stimulated by nonionic detergent, and will hydrolyze GTP nearly as efficiently as ATP. It displays MT binding that,

Fig. 5   Polarity of sea urchin egg kinesin-induced bead motility. Column-purified *S. purpuratus* kinesin was incubated with a 100-fold dilution of carboxylated latex beads (2.5% solid solution) for 10 min on ice, and then the beads were combined with astral arrays of microtubules that had been nucleated and assembled off of isolated centrosomes (C). Between 0 and 1 sec, a latex bead in solution bound to the microtubule. Images at 2.9 and 4.4 sec show this bead translocating toward the plus end of the microtubule (arrowheads). Scale bar, 1 μm. (From Porter *et al.*, 1987, Fig. 6.)

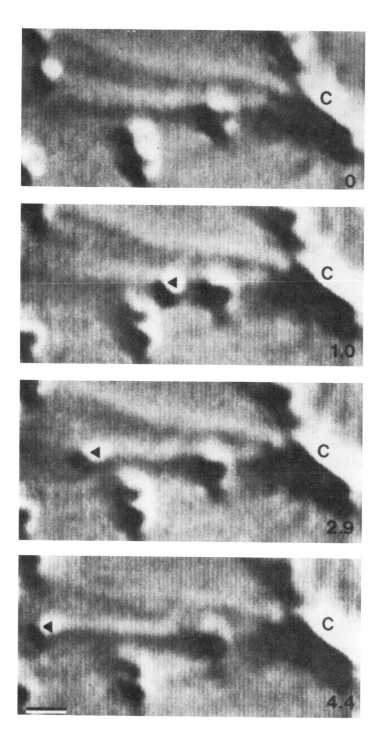

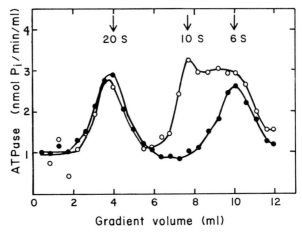

Gradient volume (ml)

Fig. 6 Sucrose density gradient analysis of sea urchin egg cytosolic extract. A cytosolic extract of
S. purpuratus eggs was fractionated on a linear sucrose density gradient (5–20% sucrose). ATPase
activity was determined in the presence (○) and absence (●) of taxol-stabilized calf brain tubulin
(0.25 mg/ml). Sedimentation values indicated were based on positions of marker proteins. (Adapted
from Collins and Vallee, 1986, Fig. 3.)

unlike kinesin, is ATP insensitive. No MT motility has been shown for the
ATPase, but Collins and Vallee (1986) have proposed that multiple MT-binding
sites may allow the protein to attach to one MT and move it along another, thus
mediating MT–MT sliding. This mechanism is similar to that proposed for
dynamin in other eukaryotes (reviewed in Vallee and Shpetner, 1990).

VI. Functions of Microtubule Motors
in Early Sea Urchin Embryos

Much is known about how axonemal dyneins, in sperm flagella and embryonic
cilia, generate forces to drive sliding between outer doublet MTs. It is less clear
how these sliding forces are translated into the bending and torsional movements
produced by intact axonemes, what the specific roles are of the various outer and
inner arm dyneins, and how they are regulated (reviewed in Porter and Johnson,
1989).

Comparatively little is known about the molecular mechanisms responsible for
MT-based movements in dividing cells. The list of candidate motors is growing
rapidly, as is our understanding of their structural and enzymatic properties *in
vitro*. However, direct evidence for roles *in vivo* will require the specific localization of these motors to cytoplasmic structures and the ability to perturb their
motility functions.

Using light microscopic immunocytochemistry, it has been found that dynein

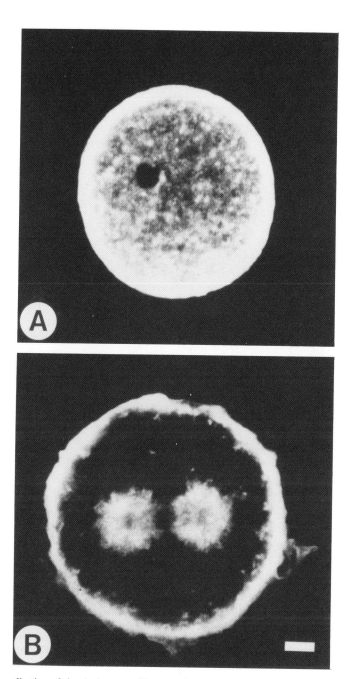

Fig. 7 Localization of dynein in sea urchin eggs visualized by immunofluorescence using anti-dynein (fragment A) serum. *Hemicentrotus pulcherrimus* embryo (A) immediately after fertilization; (B) at metaphase showing a concentration of dynein into the MA. Scale bar, 10 μm. (Adapted from Mohri *et al.*, 1976, Fig. 2.)

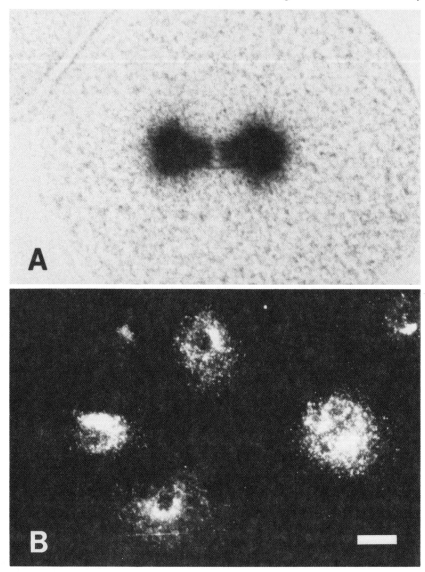

Fig. 8 Immunolocalization of kinesin in first metaphase sea urchin embryos and nonmitotic co-elomocytes using monoclonal antikinesins. (A) Peroxidase staining for kinesin in a first metaphase *L. pictus* embryo shows a concentration of staining within the mitotic spindle. (B) Fluorescent staining for kinesin shows a punctate perinuclear localization pattern in *S. droebachiensis* adult stage pha-gocytic amoebocytes (see Wright *et al.*, 1991, for details). Scale bar, 10 μm.

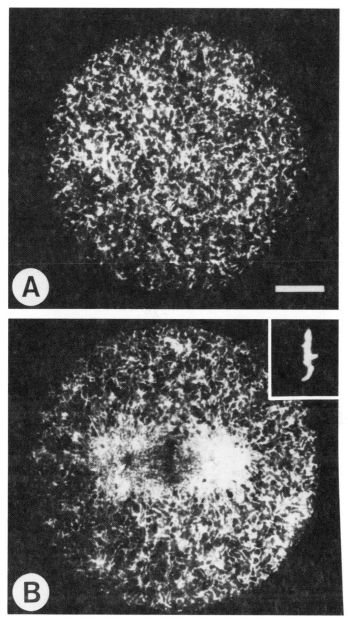

Fig. 9 Immunofluorescent localization of calsequestrin in frozen sections of first cell cycle *S. purpuratus* embryos. (A) Interphase embryo (30 min after fertilization) has a tubuloreticulum extending throughout the cytoplasm resembling the network present in the unfertilized egg. (B) Metaphase embryo showing concentration of punctate and vesicular stained elements within the MA, particularly in the asters. Inset shows Hoescht-stained chromosomes aligned on the metaphase plate. Scale bar, 10 μm. (Adapted from Henson *et al.*, 1989, Fig. 4; reproduced from the *Journal of Cell Biology*, by copyright permission of the Rockefeller University Press.)

and kinesin are concentrated in MAs of dividing sea urchin blastomeres in the early embryo (Mohri *et al.*, 1976; Scholey *et al.*, 1985; see Figs. 7 and 8A). A similar accumulation of membranes in these MAs has been documented by several investigators (Rebhun, 1960; Harris, 1962, 1975; Lee and Epel, 1983; Petzelt and Hafner, 1986; Henson *et al.*, 1989); for example, calsequestrin-containing endoplasmic reticulum (ER)-derived membranes (Oberdorf *et al.*, 1988) are strikingly concentrated in the MA (Henson *et al.*, 1989; Fig. 9).

Based on the changes in distribution of sea urchin kinesin that occur during experimental manipulation of MTs and membranes in mitotic cells, we have proposed that kinesin is associated with the ER-derived vesicles that accumulate in the MA (Wright *et al.*, 1991). This idea is supported by the observation that antikinesin stains vesicle-like structures in the cytoplasm of nonmitotic co-elomocytes (Wright *et al.*, 1991; Fig. 8B). Our current working hypothesis is that kinesin and dynein serve as plus and minus end-directed MT-based vesicle motors, respectively, in the early sea urchin embryonic MA, as well as in the interphase cytoplasm. As in other systems, then, these motors are likely to drive bidirectional membrane movements; such movements might include radial membrane transport (Hollenbeck and Swanson, 1990) and vesicle trafficking during secretion (Sinha and Wagner, 1987) and endocytosis (Bomsel *et al.*, 1990), for example.

The role of these and other MT-based motors in the complex motions associated with mitosis in the sea urchin embryo is not known. Whether cytoplasmic dynein and kinesin serve as MT-based plus and minus end-directed kinetochore motors is not known. It is possible that the 10S MT-activated ATPase may be a sea urchin "dynamin" that mediates MT–MT sliding during spindle elongation, but direct evidence for this hypothesis is lacking (discussed in Vallee *et al.*, 1990; Vallee and Shpetner, 1990). In other systems, kinesin-like motors that participate in mitosis and meiosis have been identified genetically. The rapidly dividing cells of the early sea urchin embryo should continue to be a useful system for the identification and characterization of these and other MT-based motors that are likely to participate in the various forms of MT-based motility that occur during cell division and early development.

Acknowledgments

Work in our lab is supported by grants from the NIH (GM46376-01), the American Cancer Society (#BE-46C) and March of Dimes Birth Defects Research Foundation Grant (#1-1188) to J.M.S. B.D.W. was supported by March of Dimes Birth Defects Foundation Predoctoral Graduate Research Training Fellowship #18-89-44.

References

Asai, D. J. (1986). An antiserum to the sea urchin 20 S egg dynein reacts with embryonic ciliary dynein but it does not react with the mitotic apparatus. *Dev. Biol.* **118**, 416–424.

Auclair, W., and Siegel, B. W. (1966). Cilia regeneration in the sea urchin embryo: Evidence for a pool of ciliary proteins. *Science* **154**, 913–915.

Bloom, G. S., and Vallee, R. B. (1989). Microtubule-associated proteins in the sea-urchin egg mitotic spindle. In "Mitosis: Molecules and Mechanisms" (J. Hyams and B. Brinkley, eds.), pp. 183–201. Academic Press, San Diego, California.

Bomsel, M., Parton, R., Kuznetsov, S. A., Schroer, T. A., and Gruenberg, J. (1990). Microtubule- and motor-dependent fusion *in vitro* between apical and basolateral endocytic vesicles from MDCK cells. *Cell (Cambridge, Mass.)* **62**, 719–731.

Brady, S. T., Pfister, K. K., and Bloom, G. S. (1990). A monoclonal antibody against kinesin inhibits both anterograde and retrograde fast axonal transport in squid axoplasm. *Proc. Natl. Acad. Sci. U.S.A.* **87**, 1061–1965.

Buster, D. B., and Scholey, J. M. (1991). Purification and assay of kinesin from sea urchin eggs and early embryos. *J. Cell Sci.* **14s**, 109–115.

Cohn, S. A., Ingold, A. L., and Scholey, J. M. (1989). Quantitative analysis of sea urchin egg kinesin-driven microtubule motility. *J. Biol. Chem.* **264**, 4290–4297.

Collins, C. A., and Vallee, R. B. (1986). A microtubule-activated ATPase from sea urchin eggs, distinct from cytoplasmic dynein and kinesin. *Proc. Natl. Acad. Sci. U.S.A.* **83**, 4799–4803.

Davidson, E. H., Hough-Evans, B. R., and Britten, R. J. (1982). Molecular biology of the sea urchin embryo. *Science* **217**, 17–26.

Dinsmore, J. H., and Sloboda, R. D. (1989). Microinjection of antibodies to a 62 kd mitotic apparatus protein arrests mitosis in dividing sea urchin embryos. *Cell (Cambridge, Mass.)* **57**, 127–134.

Endow, S. A., Henikoff, S., and Soler-Neidziela, L. (1990). Mediation of meiotic and early mitotic chromosome segregation in *Drosophila* by a protein related to kinesin. *Nature (London)* **345**, 81–83.

Enos, A. P., and Morris, N. R. (1990). Mutation of a gene that encodes a kinesin-like protein blocks nuclear division in *A. nidulans*. *Cell (Cambridge, Mass.)* **60**, 1019–1027.

Euteneuer, U., Koonce, M. P., Pfister, K. K., and Schliwa, M. (1988). An ATPase with properties expected for the organelle motor of the giant amoeba, *Reticulomyxa*. *Nature (London)* **332**, 176–178.

Fox, L. A., and Sale, W. S. (1987). Direction of force generated by the inner row of dynein arms on flagellar microtubules. *J. Cell Biol.* **105**, 1781–1787.

Gibbons, B. H., and Gibbons, I. R. (1972). Flagellar movement and adenosine triphosphatase activity in sea urchin sperm extracted with Triton X-100. *J. Cell Biol.* **54**, 75–97.

Gibbons, B. H., and Gibbons, I. R. (1973). The effect of partial extraction of dynein arms on the movement of reactivated sea-urchin sperm. *J. Cell Sci.* **13**, 337–357.

Gibbons, B. H., and Gibbons, I. R. (1976). Functional recombination of dynein-1 with demembranated sea urchin sperm partially extracted with KCl. *Biochem. Biophys. Res. Commun.* **73**, 1–6.

Gibbons, B. H., and Gibbons, I. R. (1979). Relationship between the latent adenosine triphosphatase state of dynein 1 and its ability to recombine functionally with KCl-extracted sea urchin sperm flagella. *J. Biol. Chem.* **254**, 197–201.

Gibbons, B. H., and Gibbons, I. R. (1987). Vanadate-sensitized cleavage of dynein heavy chains by 365 nm irradiation of demembranated sperm flagella and its effect on flagellar motility. *J. Biol. Chem.* **262**, 8354–8359.

Gibbons, B. H., Ogawa, K., and Gibbons, I. R. (1976). The effect of antidynein 1 serum on the movement of reactivated sea urchin sperm. *J. Cell Biol.* **71**, 823–831.

Gibbons, I. R. (1965). Chemical dissection of the cilia. *Arch. Biol.* **76**, 317–352.

Gibbons, I. R. (1975). The molecular basis of flagellar movement in sea urchin spermatozoa. In "Molecules and Cell Movement" (S. Inoué and R. Stephens, eds.), pp. 207–232. Raven Press, New York.

Gibbons, I. R. (1981). Cilia and flagella of eukaryotes. *J. Cell Biol.* **91**, 107s–124s.

Gibbons, I. R. (1988). Dynein ATPases as microtubule motors. *J. Biol. Chem.* **263**, 15837–15840.

Gibbons, I. R., and Gibbons, B. H. (1987). Photocleavage of dynein heavy chains. *In* "Perspectives of Biological Energy Transduction" (M. Nakao, S. Ebash, M. Morales, and E. Fleischer, eds.), pp. 107–116. Academic Press, Tokyo.

Gibbons, I. R., Fronk, E., Gibbons, B. H., and Ogawa, K. (1976). Multiple forms of dynein in sea urchin sperm flagella. *In* "Cell Motility" (R. Goldman, T. Pollard, and J. Rosenbaum, eds.), pp. 915–932. Cold Spring Harbor Lab., Cold Spring Harbor, New York.

Hafner, M., and Petzelt, C. (1987). Inhibition of mitosis by an antibody to the mitotic calcium transport system. *Nature (London)* **330**, 264–266.

Hagan, I., and Yanagida, M. (1990). Novel potential mitotic motor protein encoded by the fission yeast *cut7+* gene. *Nature (London)* **347**, 563–566.

Harris, P. (1962). Some structural aspects of the mitotic apparatus in sea urchin embryos. *J. Cell Biol.* **14**, 475–487.

Harris, P. (1975). The role of membranes in the organization of the mitotic apparatus. *Exp. Cell Res.* **94**, 409–425.

Henson, J. H., Begg, D. A., Beaulieu, S. M., Fishkind, D. J., Bonder, E. M., Terasaki, M., Lebeche, D., and Kaminer, B. (1989). A calsequestrin-like protein in the endoplasmic reticulum of the sea urchin: Localization and dynamics in the egg and first cell cycle embryo. *J. Cell Biol.* **109**, 149–161.

Hirokawa, N., Pfister, K. K., Yorifuji, H., Wagner, M. C., Brady, S. T., and Bloom, G. S. (1989). Submolecular domains of bovine brain kinesin identified by electron microscopy and monoclonal antibody decoration. *Cell (Cambridge, Mass.)* **56**, 867–878.

Hollenbeck, P. J., and Swanson, J. A. (1990). Radial extension of macrophage tubular lysosomes supported by kinesin. *Nature (London)* **346**, 864–866.

Ingold, A. L., Cohn, S. A., and Scholey, J. M. (1988). Inhibition of kinesin-driven microtubule motility by monoclonal antibodies to kinesin heavy chains. *J. Cell Biol.* **107**, 2657–2667.

Johnson, C. S., Buster, D., and Scholey, J. M. (1990). Light chains of sea urchin kinesin identified by immunoadsorption. *Cell Motil. Cytoskel.* **16**, 204–213.

Kiehart, D. P. (1982). Microinjection of echinoderm eggs: Apparatus and procedures. *Methods Cell Biol.* **25b**, 13–31.

Kiehart, D. P., Mabuchi, I., and Inoué, S. (1982). Evidence that myosin does not contribute to force production in chromosome movement. *J. Cell Biol.* **94**, 165–178.

Kimura, I. (1977). Ciliary dynein from sea urchin embryos. *J. Biochem. (Tokyo)* **81**, 715–720.

Kirschner, M., and Mitchison, T. (1986). Beyond self-assembly: From microtubules to morphogenesis. *Cell (Cambridge, Mass.)* **45**, 329–342.

Koonce, M. P., Tong, J., Euteneuer, U., and Schliwa, M. (1987). Active sliding between cytoplasmic microtubules. *Nature (London)* **328**, 737–739.

Lee, H. C., and Epel, D. (1983). Changes in intracellular acidic compartments in sea urchin eggs after activation. *Dev. Biol.* **98**, 446–454.

Lye, R. J., Porter, M. E., Scholey, J. M., and McIntosh, J. R. (1987). Identification of a microtubule-based cytoplasmic motor in the nematode. *C. elegans. Cell (Cambridge, Mass.)* **51**, 309–318.

Lye, R. J., Pfarr, C. M., and Porter, M. E. (1989). Cytoplasmic dynein and microtubule translocators. *In* "Kinesin, Dynein, and Microtubule Dynamics" (F. Warner and J. McIntosh, eds.), Cell Movement, Vol. 2, pp. 141–154. Liss, New York.

Mabuchi, I., and Okuno, M. (1977). The effect of myosin antibody on the division of starfish blastomeres. *J. Cell Biol.* **74**, 251–263.

Masuda, H., McDonald, K. L., and Cande, W. Z. (1988). The mechanism of anaphase spindle elongation: Uncoupling of tubulin incorporation and microtubule sliding during *in vitro* spindle reactivation. *J. Cell Biol.* **107**, 623–633.

McDonald, H. B., and Goldstein, L. S. B. (1990). Identification and characterization of gene encoding a kinesin-like protein in *Drosophila. Cell (Cambridge, Mass.)* **61**, 991–1000.

McDonald, H. B., Stewart, R. J., and Goldstein, L. S. B. (1990). The kinesin-like *ncd* protein of *Drosophila* is a minus end-directed microtubule motor. *Cell (Cambridge, Mass.)* **63**, 1159–1165.

McIntosh, J. R. (1983). The centrosome as organizer of the cytoskeleton. *Mod. Cell Biol.* **2**, 115–142.

McIntosh, J. R., and Porter, M. E. (1989). Enzymes for microtubule-dependent motility. *J. Biol. Chem.* **264**, 6001–6004.

Meluh, P. B., and Rose, M. D. (1990). *KAR3*, a kinesin-related gene required for yeast nuclear fusion. *Cell (Cambridge, Mass.)* **60**, 1029–1041.

Mocz, G., Tang, W. J. Y., and Gibbons, I. R. (1988). A map of photolytic and tryptic cleavage sites in the β heavy chain of outer arm dynein from sea urchin sperm flagella. *J. Cell Biol.* **106**, 1607–1614.

Mohri, H. Mohri, T., Mabuchi, I., Yazaki, I., Sakai, H., and Ogawa, Y. (1976). Localization of dynein in sea urchin eggs during cleavage. *Dev., Growth Differ.* **18**, 391–398.

Oberdorf, J. A., Lebeche, D., Head, J. F., and Kaminer, B. (1988). Identification of a calsequestrin-like protein from sea urchin eggs. *J. Biol. Chem.* **263**, 6806–6809.

Ogawa, K., Mohri, T., and Mohri, H. (1977). Identification of dynein as the outer arms of sea urchin sperm axonemes. *Proc. Natl. Acad. Sci. U.S.A.* **74**, 5006–5010.

Ogawa, K., Negishi, S., and Obika, M. (1982). Immunological dissimilarity in protein component (dynein 1) between outer and inner arms within sea urchin sperm axonemes. *J. Cell Biol.* **92**, 706–713.

Paschal, B. M., and Vallee, R. B. (1987). Retrograde transport by the microtubule-associated protein MAP 1C. *Nature (London)* **330**, 181–183.

Paschal, B. M., King, S. M., Moss, A. G., Collins, C. A., Vallee, R. B., and Witman, G. B. (1987). Isolated flagellar outer arm dynein translocates brain microtubules *in vitro*. *Nature (London)* **330**, 672–674.

Petzelt, C., and Hafner, M. (1986). Visualization of the Ca^{2+}-transport system of the mitotic apparatus of sea urchin eggs with a monoclonal antibody. *Proc. Natl. Acad. Sci. U.S.A.* **83**, 1719–1722.

Pfarr, C. M. Coue, M., Grissom, P. M., Hays, T. S., Porter, M. E., and McIntosh, J. R. (1990). Cytoplasmic dynein is localized to kinetochores during mitosis. *Nature (London)* **345**, 263–265.

Piperno, G. (1984). Monoclonal antibodies to dynein subunits reveal the existence of cytoplasmic antigens in sea urchin egg. *J. Cell Biol.* **98**, 1842–1850.

Porter, M. E., and Johnson, K. A. (1989). Dynein structure and function. *Annu. Rev. Cell Biol.* **5**, 119–151.

Porter, M. E., Scholey, J. M., Stemple, D. L., Vigers, G. P. A., Vale, R. D., Sheetz, M. P., and McIntosh, J. R. (1987). Characterization of the microtubule movement produced by sea urchin egg kinesin. *J. Biol. Chem.* **262**, 2794–2802.

Porter, M. E., Grissom, P. M., Scholey, J. M., Salmon, E. D., and McIntosh, J. R. (1988). Dynein isoforms in sea urchin eggs. *J. Biol. Chem.* **263**, 6759–6771.

Pratt, M. M. (1989). Dyneins in sea urchin eggs and nervous tissue. *In* "Kinesin, Dynein, and Microtubule Dynamics" (F. Warner and J. McIntosh, eds.), Cell Movement, Vol. 2, pp. 125–140. Liss, New York.

Pryer, N. K., Wadsworth, P., and Salmon, E. D. (1986). Polarized microtubule gliding and particle saltations produced by soluble factors from sea urchin eggs and embryos. *Cell Motil. Cytoskel.* **6**, 537–548.

Rebhun, L. I. (1960). Aster-associated particles in the cleavage of marine invertebrate eggs. *Ann. N.Y. Acad. Sci.* **90**, 357–380.

Sale, W. S., and Fox, L. A. (1986). High molecular weight polypeptides of the inner dynein arms in sea urchin sperm tails. *J. Cell Biol.* **103**, 4a.

Sale, W. S., Goodenough, U. W., and Heuser, J. E. (1985). The substructure of isolated and *in situ* outer dynein arms of sea urchin sperm flagella. *J. Cell Biol.* **101**, 1400–1412.

Sale, W. S., Fox, L. A., and Milgram, S. L. (1989). Composition and organization of the inner row dynein arms. *In* "The Dynein ATPases" (F. Warner, P. Satir, and I. Gibbons, eds.), pp. 89–102. Liss, New York.

Saxton, W. M., Hicks, J., Goldstein, L. S. B., and Raff, E. C. (1991). Kinesin heavy chain mutants cause neuromuscular malfunction and lethality in *Drosophila* larvae. *Cell (Cambridge, Mass.)* **64,** 1093–1102.

Schnapp, B. J., and Reese, T. S. (1989). Dynein is the motor for retrograde axonal transport of organelles. *Proc. Natl. Acad. Sci. U.S.A.* **86,** 1548–1552.

Scholey, J. M., Neighbors, B., McIntosh, J. R., and Salmon, E. D. (1984). Isolation of microtubules and a dynein-like MgATPase from unfertilized sea urchin eggs. *J. Biol. Chem.* **259,** 6516–6525.

Scholey, J. M., Porter, M. E., Grissom, P. M., and McIntosh, J. R. (1985). Identification of kinesin in sea urchin eggs, and evidence for its localization in the mitotic spindle. *Nature (London)* **318,** 483–486.

Scholey, J. M., Heuser, J., Yang, J. T., and Goldstein, L. S. B. (1989). Identification of globular mechanochemical heads of kinesin. *Nature (London)* **338,** 355–357.

Schroeder, T. E., ed. (1986). "Methods in Cell Biology: Echinoderm Gametes and Embryos," Vol. 27. Academic Press, Orlando, Florida.

Schroer, T. A., Schnapp, B. J., Reese, T. S., and Sheetz, M. P. (1988). The role of kinesin and other soluble factors in organelle movement along microtubules. *J. Cell Biol.* **107,** 1785–1792.

Schroer, T. A., Steuer, E. R., and Sheetz, M. P. (1989). Cytoplasmic dynein is a minus end-directed motor for membranous organelles. *Cell (Cambridge, Mass.)* **56,** 937–946.

Shpetner, H. S., and Vallee, R. B. (1989). Identification of dynamin, a novel mechanochemical enzyme that mediates interactions between microtubules. *Cell (Cambridge, Mass.)* **59,** 421–432.

Sinha, S., and Wagner, D. D. (1987). Intact microtubules are necessary for complete processing, storage and regulated secretion of von Willebrand factor by endothelial cells. *Eur. J. Cell Biol.* **43,** 377–383.

Steuer, E. R., Wordeman, L., Schroer, T. A., and Sheetz, M. P. (1990). Localization of cytoplasmic dynein to mitotic spindles and kinetochores. *Nature (London)* **345,** 266–268.

Summers, K. E., and Gibbons, I. R. (1973). Effects of trypsin digestion on flagellar structures and their relationship to motility. *J. Cell Biol.* **58,** 618–629.

Vale, R. D. (1987). Intracellular transport using microtubule-based motors. *Annu. Rev. Cell Biol.* **3,** 347–378.

Vale, R. D. (1990). Microtubule-based motor proteins. *Curr. Opin. Cell Biol.* **2,** 15–22.

Vale, R. D., and Goldstein, L. S. B. (1990). One motor, many tails: An expanding repertoire of force-generating enzymes. *Cell (Cambridge, Mass.)* **60,** 883–885.

Vale, R. D., Reese, T. S., and Sheetz, M. P. (1985a). Identification of a novel force generating protein, kinesin, involved in microtubule-based motility. *Cell (Cambridge, Mass.)* **42,** 39–50.

Vale, R. D., Schnapp, B. J., Mitchison, T., Steuer, E., Reese, T. S., and Sheetz, M. P. (1985b). Different axoplasmic proteins generate movement in opposite directions along microtubules *in vitro*. *Cell (Cambridge, Mass.)* **43,** 623–632.

Vale, R. D., Soll, D. R., and Gibbons, I. R. (1989). One-dimensional diffusion of microtubules bound to flagellar dynein. *Cell (Cambridge, Mass.)* **59,** 915–925.

Vallee, R. B., and Shpetner, H. S. (1990). Motor proteins of cytoplasmic microtubules. *Annu. Rev. Biochem.* **59,** 909–932.

Vallee, R. B., Shpetner, H. S., and Paschal, B. M. (1990). Potential roles of microtubule-associated motor molecules in cell division. *Ann. N.Y. Acad. Sci.* **582,** 99–107.

Walker, R. A., Salmon, E. D., and Endow, S. A. (1990). The *Drosophila claret* segregation protein is a minus-end directed motor molecule. *Nature (London)* **347,** 780–782.

Warner, F. D., Satir, P., and Gibbons, I. R., eds. (1989). "Cell Movement," Vol. 1. Liss, New York.

Weisenberg, R. C., and Taylor, E. W. (1968). Studies on ATPase activity of sea urchin eggs and isolated mitotic apparatus. *Exp. Cell Res.* **53**, 372–384.

Wright, B. D., Henson, J. H., Wedaman, K. P., Willy, P. J., Morand, J. N., and Scholey, J. M. (1991). Subcellular localization and sequence of sea urchin kinesin heavy chain: Evidence for its association with membranes in the mitotic apparatus and interphase cytoplasm. *J. Cell Biol.* **113**, 817–833.

Zhang, P., Knowles, B. A., Goldstein, L. S. B., and Hawley, R. S. (1990). A kinesin-like protein required for distributive chromosome segregation in *Drosophila*. *Cell (Cambridge, Mass.)* **62**, 1053–1062.

NOTE ADDED IN PROOF: The sea urchin kinesin heavy chain sequence (Wright *et al.*, 1991) is now available from EMBL/GenBank/DDBJ under accession number X56844. Sea urchin axonemal dyein β-heavy chain sequences have recently been obtained (Gibbons, I. R., Gibbons, B. H., Mocz, G., and Asai, D. J. (1991). *Nature (London)* **352**, 640–643; Ogawa, K. (1991). *Nature (London)* **352**, 643–645) and appear in the EMBL/GenBank/DDBJ databases under accession numbers X59603 and D01021, respectively.

6

Assembly of the Intestinal Brush Border Cytoskeleton

Matthew B. Heintzelman and Mark S. Mooseker
Department of Biology
Yale University
New Haven, Connecticut 06511

I. Introduction

A variety of epithelial cell types display microvilli (MV) at their apical surface, but only a small subset of these, including the intestinal absorptive cell and the kidney proximal tubule cell, display what has been appropriately termed a brush border (BB). The apical surface of these cells is covered with a dense population of uniform-length MV, the most apparent function of which is to increase the cell's surface area for absorptive processes. The BB is not, however, a static

Current Topics in Developmental Biology, Vol. 26
Copyright © 1992 by Academic Press, Inc. All rights of reproduction in any form reserved.

structure but can undergo some striking changes in response to a variety of stimuli, including feeding and fasting (e.g., Misch *et al.*, 1980), or exposure to lectins (Draaijer *et al.*, 1989; Weinmann *et al.*, 1989). Furthermore, the major protein components of the BB cytoskeleton are themselves undergoing constant turnover (Stidwell *et al.*, 1984) and yet the cell maintains the structure of this highly ordered domain. Obviously, the assembly and maintenance of such a regular cytoskeletal array as the BB is no small feat of ultrastructural engineering. As such, it provides a multifaceted experimental model for studying cytoskeletal assembly in terms of both protein–protein and protein–membrane interactions.

The objective of this article is to explore our current understanding of BB assembly. Although the complete story has yet to be written, experimental and descriptive studies of BB and MV assembly from a number of laboratories have yielded some insight into the mechanisms of cytoskeletal assembly. A variety of experimental systems are being used and include the developing embryonic epithelium of both intestine and kidney, the proliferative epithelium of the adult intestine, a variety of cell lines derived from neoplastic intestine (HT-29, Caco2, T84) and kidney (LLC-PK$_1$), and, more recently, invertebrate tissues including the retina (rhabdome) and midgut of *Drosophila* and other arthropods. In addition, assembly mechanisms are also being studied by means of transfection and microinjection experiments in which brush borderless cell types are used as models to explore the potential of BB proteins to organize cytoarchitecture. Together, this combination of approaches has yielded, and should continue to yield, important information regarding the processes by which a cell can construct a complex cytoskeletal array such as the brush border.

II. The Brush Border Cytoskeleton

The BB has been amenable to careful analysis since it was first isolated by Miller and Crane (1961), who sought to localize the disaccharidases of the small intestine. Subsequent analyses of the isolated BB, primarily from chicken intestine, began to explore the cytoskeletal constituents of the BB, the primary component of which was first identified as actin by Tilney and Mooseker (1971). Since that time, the isolated BB has become increasingly well characterized and now serves as a model system for cytoskeletal–membrane assembly and interactions (Mooseker, 1985; Coudrier *et al.*, 1988; Louvard, 1989).

Morphologically, the BB can be considered as two structurally distinct regions—the MV domain and the terminal web (TW) domain. The MV domain consists of the 1- to 2-μm long, finger-like MV that contain at their core a uniformly polarized bundle of actin filaments and a host of actin-binding proteins. These cores of actin filaments continue downward into the subjacent cytoplasm as rootlets and establish a TW domain. This region is further embellished by a set of proteins that interconnect the MV rootlets with each other and with the intercellular junctional complexes (Fig. 1).

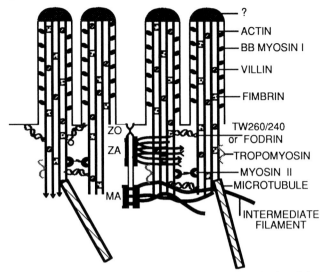

Fig. 1 A simplified model depicting the molecular and structural organization of the brush border cytoskeleton. Some BB constituents known to exist in the MV domain (e.g., p80) or present in the TW domain (e.g., p36/p10), but still lacking structural assignment, have not been included. Similarly, it is thought that Z-protein might be at least one constituent of the dense cap at the tips of MV (see text for details.)

A. The Microvillus Core

In the MV core, the actin filaments are held together by two major proteins, fimbrin and villin. Fimbrin is a 68-kDa polypeptide that has been identified in a variety of cell types, where it is typically associated with uniformly polarized arrays of F-actin such as membrane ruffles, filopodia, and stereocilia (Bretscher and Weber, 1980). *In vitro*, fimbrin can interact with actin to form highly ordered bundles of actin (Glenney *et al.*, 1981) akin to those seen *in vivo* and is thus thought to be the major bundling protein of the MV core.

The primary structure of fimbrin has revealed a 10-kDa, N-terminal head piece responsible for calcium binding, and a 58-kDa core containing tandemly repeated actin-binding domains similar to those seen in the family of actin gelation proteins (e.g., α-actinin, β-spectrin, dystrophin; see de Arruda *et al.*, 1990). Furthermore, the discovery of fimbrin as a homolog of L-plastin, a leukocyte-specific protein of unknown function that is phosphorylated on interleukin I activation (de Arruda *et al.*, 1990), raises the possibility that the calcium- or actin-binding activities of fimbrin might themselves be regulated by phosphorylation. Understanding the regulation of the functional domains of fimbrin and the significance of phosphorylated fimbrin isoforms awaits further biochemical characterization of both fimbrin and the plastins.

Villin is a 95-kDa protein found almost exclusively in BB-containing cell types and in certain cell types derived from the embryonic gut (Robine et al., 1985). Villin is a multifunctional protein that exhibits a variety of calcium-dependent actin-binding properties that will be summarized only briefly here, as detailed treatments of the structural and functional properties of villin are in the literature (Mooseker, 1985; Louvard, 1989; Friederich et al., 1990). In the absence of calcium, villin will bundle F-actin, although not into the ordered arrays seen with fimbrin alone or with fimbrin and villin together. In micromolar calcium, villin acts as a barbed-end capper of F-actin, and at calcium concentrations above 5 μM will sever actin filaments. In addition, villin will interact with G-actin in micromolar calcium to facilitate the nucleation phase of actin assembly. Obviously, villin could play a significant role in the assembly and maintenance of the BB cytoskeleton, and has, accordingly, been given much attention in the study of BB and MV assembly.

It is apparent from standard electron microscopy (EM) on isolated BBs that the core of actin filaments in the MV is connected to the MV membrane by a helically arranged set of cross-bridges (Millington and Finean, 1962; Mooseker and Tilney, 1975; Matsudaira and Burgess, 1979). These ridges are composed of a 110-kDa calmodulin (CM)-binding protein that, together with its bound CM subunits, has been called brush border myosin I (BB myosin I; for references, see Coluccio and Bretscher, 1989). As its name indicates, this protein is a member of the single-headed myosin family (for review, see Pollard et al., 1991) and has demonstrated mechanochemical properties (Mooseker and Coleman, 1989; Mooseker et al., 1989; Collins et al., 1990). The role of BB myosin I in the enterocyte is unknown, but the majority of speculation suggests that it may be involved in the movement of membrane elements (Drenckhahn and Dermietzel, 1988; Mooseker and Coleman, 1989; Shibayama et al., 1987).

Villin, fimbrin, and BB myosin I represent the major proteins, with known structural correlates, of the MV core. An 80-kDa protein, termed p80 or ezrin, has been identified as a minor component of the MV core and is known to be a tyrosine kinase substrate (Bretscher, 1983; Gould et al., 1986). Neither its structural nor functional role in the MV has yet been determined. Z-protein, a constituent of the myofibrillar Z-line, has also been reported to exist in the BB both at the junctional complexes and at the tips of microvilli (Ohashi and Maruyama, 1989). The immunocytochemical localization of this component has not yet been verified by cell fractionation biochemistry. Still other components are likely to exist in the MV but have not yet been identified due to their low abundance or loss during BB isolation. Nevertheless, characterization of these elements will be important in understanding the structure and function of the BB as a whole.

B. The Terminal Web

The actin cores of the MV terminate as rootlets in the TW domain of the BB. Villin and fimbrin remain associated with the actin cores at this level and, although low levels of immunogold labeling for BB myosin I have been seen in the TW domain, it is uncertain whether it is associated with the actin core, with vesicular elements in the TW region, or possibly with both (Drenckhahn and Dermietzel, 1988; Heintzelman and Mooseker, 1990a). Two major classes of proteins, myosin II and nonerythroid spectrin, form an interlacing network connecting the MV rootlets with each other (Hirokawa *et al.*, 1982, 1983; Glenney and Glenney, 1983). The organization of this domain is elegantly displayed in the quick-freeze, deep-etch preparations of Hirokawa and Heuser (1981). Bipolar filaments of myosin are oriented perpendicular to the rootlet axes and are more predominant in the lower half of the TW (Hirokawa *et al.*, 1983; Pearl *et al.*, 1984; Mooseker *et al.*, 1984). Fodrin is found in the TW of mammals and the chicken TW contains both fodrin and, more predominantly, a chicken BB-specific spectrin, TW260/240 (Hirokawa *et al.*, 1983; Glenney and Glenney, 1983; Glenney *et al.*, 1982; Carboni *et al.*, 1987). These molecules exist along the whole length of the rootlets and can be seen to connect not only the rootlets with each other but also to the overlying cell membrane, to vesicular elements in the TW, and to the local population of intermediate filaments (IF) (Hirokawa *et al.*, 1983).

Tropomyosin (TM) is also located in the TW region, presumably located along the length of the actin core rootlets (Mooseker, 1976; Bretscher and Weber, 1978; Drenckhahn and Gröschel-Stewart, 1980). *In vitro,* TM can displace both villin and fimbrin, but not BB myosin I, from MV cores (Burgess *et al.*, 1987). Why TM, *in vivo,* is associated only with the TW rootlet and not with the entire MV actin core is unclear although its TW localization may be influenced by another TW protein, caldesmon. Caldesmon is known to enhance the interaction of TM with actin and, although the ultrastructural localization of caldesmon has not been documented, its distribution coincident with that of TM in other cells types would suggest that caldesmon is associated with TM on the rootlets (Bretscher and Lynch, 1985).

Microtubules are another component of the TW domain and have been visualized oriented with the long axis of the cell in the mouse enterocyte (Sandoz and Laine, 1985; Hagen *et al.*, 1987). As with the MV domain, other proteins such as calpactin I (p36/p10) are found in the TW but lack any obvious structural correlate (Gerke and Weber, 1984; Greenberg *et al.*, 1984).

The cytoskeleton of the TW domain also intermingles with elements of the intercellular junctions. Specifically, both the zonulae occludens (ZO) and the zonula adherans (ZA) have a filamentous actin network associated with them (Madara, 1987; Hirokawa *et al.*, 1982; Keller and Mooseker, 1982). In isolated BBs, contraction of the ZA ring results in the splaying out of microvilli (Rodewald *et al.*, 1976; Keller and Mooseker, 1982; Burgess, 1982). *In situ,* this contraction may also be associated with a loosening of the ZO and coincident increases in paracellular permeability (Keller and Mooseker, 1982; Madara, 1989). It is apparent, then, that the BB cytoskeleton is intimately tied into the overall cytoskeletal network of the cell, not only with the junctional complexes but with MT and IF systems as well. Furthermore, the cytoskeletal proteins actin, villin, fimbrin, and BB myosin I are present along the basolateral membrane of the intestinal cell and may, in some way, contribute of the organization of that cellular domain as well (Heintzelman and Mooseker, 1990a). Although a brush border is not present along the basolateral domain, numerous membrane infold-ings or plicae are present and may represent the modification of the basolateral domain for increasing surface area. It appears, then, that the enterocyte may be using a similar set of building blocks to establish domains of varied morphology with the common goal of increasing surface area.

C. Variations in Brush Border Organization

To date, the small intestinal BB of the chicken and some mammalian species (mouse, human) has been far better characterized than other BB systems and will thus be the major focus of this article. These two systems are quite similar in their structural and biochemical organization and suggest that a conserved BB blueprint exists. Other BB-containing cell types show some differences, at least in structural organization, but this is generally limited to MV dimensions and packing, and the complexity of the TW domain (Fig. 2). For example, cells of the kidney proximal tubule generally have longer, thinner MV than those of the intestine, but lack an extensive TW domain (Maunsbach, 1973). However, the TW domain of the kidney contains a large population of vesicular elements, reflecting the high levels of endocytosis occurring in the proximal tubule (Rod-man et al., 1986). Colonic cells generally have shorter MV than those of the small intestine, but the TW is more elaborate as the MV rootlets are longer than in small intestine and IFs are abundant in impressive amounts (Carboni et al., 1987). The rhabdome, the BB-like array of MV that forms the light absorptive surface of the arthropod retina, has 1- to 1.5-μm long, but very tightly packed, MV complete with actin cores and core–membrane cross-linkers (e.g., Blest et al., 1982; Arikawa et al., 1990). It is possible that other biochemical differences will be found when each of these systems is better characterized.

Fig. 2 Variations on a theme. Transmission electron micrographs of intestinal BBs from the duo-dena of chicken (a) and rat (b), and the midguts of *Manduca* (c) and *Drosophila* (d) larvae show similarities and differences in BB architecture. The MV of chick and rat intestinal BBs are of similar length (1.5 μm), but the TW domain of the chick BB is more extensive. The BB in the midgut of *Manduca* has 9- to 10-μm long MV but a poorly defined TW domain. Brush borders in the midgut of *Drosophila* contain 3-μm long MV and have a thin, but well-defined, TW region and zone of exclusion. (a) and (b) are the same magnification (bar = 1 μm). Bar in (c) = 2 μm; bar in (d) = 1 μm.

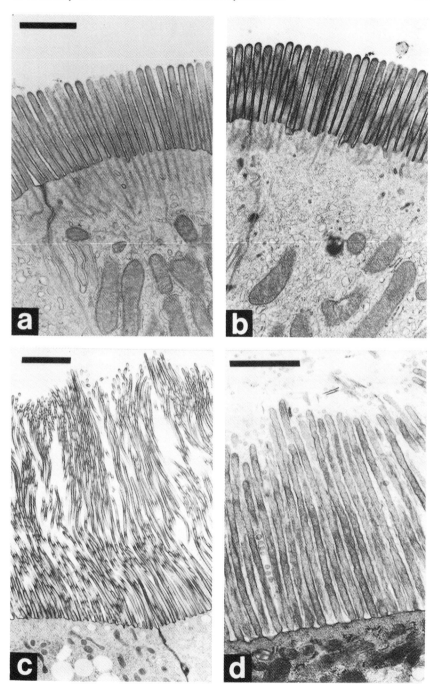

COLORADO COLLEGE LIBRARY
COLORADO SPRINGS
COLORADO

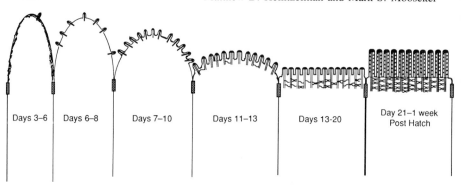

| Days 3–6 | Days 6–8 | Days 7–10 | Days 11–13 | Days 13-20 | Day 21–1 week Post Hatch |

Fig. 3 A model depicting the major events occurring during intestinal BB assembly in the chicken embryo. *Days 3–6:* The apical domain of the cell balloons out into the lumen of the gut, its cortex rich in F-actin and villin. Only an occasional MV is present. Fimbrin and BB myosin I are expressed, but are diffusely distributed throughout the cytoplasm. *Days 6–8:* The number of MV gradually increases, their appearance being a spatially random process. *Days 7–10:* Cell apices begin to flatten concomitant with increases in MV density. Fimbrin begins its redistribution to the apical domain. *Days 11–13:* The cells continue to flatten and MV density and regularity has increased. F-Actin, villin, and fimbrin are all apically localized while BB myosin I remains diffusely localized. The MV extend short rootlets into the presumptive TW domain, which contains occasional myosin II and fodrin interrootlet cross-linkers. *Days 13–20:* MV density continues to increase and the TW undergoes further organization as the density of interrootlet cross-linkers, primarily fodrin and myosin II, increases. By day 18, BB myosin I begins its redistribution to the cell periphery. By day 19, TW260/240 is expressed in the TW and becomes the predominate spectrin isoform in the TW. *Day 21–1 week posthatch:* A 3- to 4-fold increase in MV length occurs, the TW undergoes final stratification, and BB myosin I becomes localized to the MV domain.

III. Assembly of the Intestinal Brush Border Cytoskeleton during Embryogenesis: Morphological Studies

Although a number of systems are available for the study of BB structure and assembly, our best source of information about the actual assembly process has come from the embryonic intestine of both chickens and mammals. Unlike epithelial cell renewal and BB assembly in the adult intestine, which occurs in only a few days (Cheng and Leblond, 1974), BB assembly in the embryonic gut, particularly in the chicken, occurs gradually over the course of a few weeks. Consequently, our ability to resolve individual assembly events is quite good using both standard morphological and immunocytochemical techniques (Fig. 3).

A. Chick Intestinal Embryogenesis

A wealth of morphological studies has given us a clear description of epithelial differentiation in the embryonic chicken gut. The earliest studies focused on the

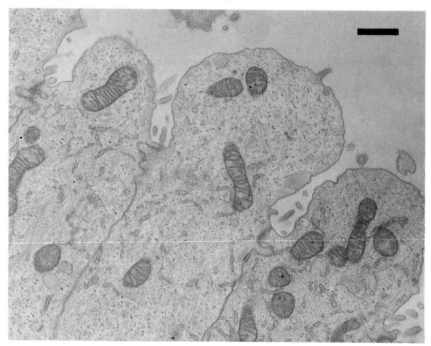

Fig. 4 Transmission electron micrograph of an intestinal epithelial cell from a 7.5-day chick embryo. The undifferentiated apical domain bulges into the lumen of the gut despite the presence of intact junctional complexes. Only an occasional, short MV is present on the surface. Bar = 1 μm.

gross changes in epithelial architecture, including the formation of villi from the embryonic previllous ridges (Hilton, 1902; Coulombre and Coulombre, 1958; Grey, 1972; Burgess, 1975). Higher resolution studies, using techniques in scanning and transmission electron microscopy, have focused on the ultrastructural changes resulting in the assembly of the BB cytoskeleton itself.

Scanning EM studies (Noda, 1981; Heintzelman and Mooseker, 1990b) reveal that the embryonic intestinal epithelium (days 6–10) lacks any structure resembling the adult brush border. The apical surfaces of these early enterocytes, resembling partially deflated balloons, bulge into the lumen of the gut, despite the presence of well-defined junctional complexes (Fig. 4). Their surface is studded with a small population of short and irregular microvillus-like processes. The next several days of development are evidenced by an increase in the density and regularity of these processes and with a coincident flattening of the cell surface. Thus, by days 12–13, the cells are covered with such a dense population of MV that the individual cell boundaries become indistinguishable, as in the adult. Increases in MV density do continue, however, and a dramatic three- to fourfold increase in MV length occurs around the time of hatching (Stidwell and Burgess, 1986).

Taken together, transmission EM studies (Overton and Shoup, 1964; Chambers and Grey, 1979; Heintzelman and Mooseker, 1990b) reveal that the balloon-like apices of the early enterocytes contain a cortical meshwork of actin. The early MV contain a loosely organized array of actin filaments, primarily oriented with the long axis of the MV and initially associated with the MV membrane. With time, the MV core of actin filaments becomes more organized (at about day 10) as more filaments are added. By embryonic days 12–13, the MV core is well organized and begins to extend into the subjacent cytoplasm as a short rootlet. The rootlets continue a gradual elongation until late in embryogenesis, when their length stabilizes at about 0.5 μm (Chambers and Grey, 1979; Takemura *et al.*, 1988).

Using quick-freeze, deep-etch electron microscopy, Takemura *et al.* (1988) described the assembly of the terminal web domain. By embryonic day 13, when most MV of the cell had short (0.2 μm) rootlets, some cross-linking elements were evident in the interrootlet zone. In addition, an intermediate filament network was already present below the level of the actin rootlets. The complexity of TW cross-linker organization increased with further development, so that by day 19 the TW organization was approaching the morphological organization seen in the adult in that a well-defined cortical zone of exclusion was evident. Further changes leading to a stratification of the TW domain into a denser bottom layer and a less dense upper layer were finally established 2–3 days after hatching, thus completing the adult TW organization (Chambers and Grey, 1979).

It is apparent that the increasing complexity of the TW domain correlates with the increasing regularity of the surface microvilli. That is, many early MV are oriented at an angle to the cell surface, but their regular, vertical arrangement proceeds with TW elaboration. In addition, TW organization is completed just prior to posthatch MV elongation (Stidwell and Burgess, 1986; Takemura *et al.*, 1988), and it has been suggested (Takemura *et al.*, 1988) that the TW may be necessary as a structural anchor for the rapid elongation of MV. However, additional roles for this TW elaboration are likely because other BB systems, such as the kidney proximal tubule, are able to assemble long MV without an extensive TW domain.

B. Mammalian Intestinal Embryogenesis

Less attention has been paid to the details of BB cytoskeletal assembly in mammalian species compared to the chicken. Nevertheless, the overall pattern of BB assembly seen in rat, mouse, and human fetal enterocytes is very similar to that seen in the avian system, the primary difference being one of developmental timing. Thus, while the chick embryo gradually assembles its BB cytoskeleton over the course of its 3-week incubation, the rat and mouse hastily assemble their BB in the last 4–5 days of their 3-week gestation, and then further modify it postnatally in response to nursing and weaning (Overton, 1965). In the human and the guinea pig, however, structural assembly of both MV and TW domains of the BB are near completion well before birth (Bailey *et al.*, 1984; Kelley, 1973).

In both rat and mouse, no evidence of BB formation is evident in the 15-day embryo, the pseudostratified epithelium of the gut resembling that of a 5- to 6-day chick embryo (Mathan *et al.*, 1976; Rochette-Egly and Haffen, 1987; Ezzell *et al.*, 1989). By 17 days in the rat, a few MV have appeared, but the morphology is still comparable to that of an early stage chick embryo. Not until 19 days in the rat (16–17 days in the mouse) is the simple epithelium established and covered with a uniform lawn of short MV comparable to a 13- to 14-day chick embryo (Mathan *et al.*, 1976; Ezzell *et al.*, 1989). Overton (1965) looked specifically at MV development in the mouse embryo and noted little MV growth until 18 days of development, when a threefold increase in length occurred. This was followed, after birth, by a gradual increase in length during nursing. At the time of weaning, MV length increased another 20%.

Brush border assembly in the human intestine appears to follow the same morphological patterns described above, but most of the major events in BB assembly have occurred by 22 weeks of development. The exact stage of development depends not only on the age of the fetal intestine examined, but also on the region of tissue sampled, because a proximal–distal gradient of development exists in the gut (e.g., see Rochette-Egly *et al.*, 1988). Thus, the 8-week human small intestine, although still a pseudostratified epithelium, has a well-developed lawn of 1-μm long MV (Kelley, 1973). By 9–10 weeks, the epithelium is columnar, and by 15 weeks the microvilli have elongated further and a thin TW zone of exclusion is evident (Moxey and Trier, 1979). By 20–22 weeks, the small intestine BB resembles that of the adult (Moxey and Trier, 1979).

The human colon goes through similar stages, although the timing lags behind that of the small bowel. Thus, colonic enterocytes of an 8-week embryo have a bulbous apical surface with a few MV, similar to that seen in a 9- to 10-day chick embryo (Lacroix *et al.*, 1984). By 10.5–11 weeks the apical surface begins to flatten, but MV density is still not great (cf. days 11–12 chick embryo) and no TW is yet present (Bell and Williams, 1982). By 16 weeks, however, a well-defined BB is present.

It is apparent from the above descriptions that a conserved blueprint for intestinal BB assembly appears to exist. That is, the same basic steps seem to occur in all the species examined thus far, the primary difference being one of developmental timing and the actual rate of assembly. Of the available systems, the embryonic chick gut seems to be the optimal system as tissue procurement is easier and BB assembly occurs more gradually than in the rodent models.

IV. Patterns of Brush Border Protein Expression during Intestinal Embryogenesis

A further level of resolution into the mechanisms of BB cytoskeletal assembly is achieved using immunochemical techniques. We can begin to correlate the pattern of expression of BB proteins with structural changes occurring during development and, in so doing, further refine our ideas regarding cytoskeletal assembly at the molecular level. Just as the embryonic chicken gut has been more thoroughly described in morphological studies than have the other systems, so too is our understanding of BB protein expression more advanced in the chick than in mammalian systems. However, recent information regarding mammalian brush border expression does suggest that similar patterns of expression do occur—a finding one might expect given the morphological similarities between avian and mammalian systems.

A. Immunochemical and Localization Studies of MV Core Protein Expression in the Chick Embryo

1. The Initiation of BB Assembly

Given that actin is at the core of the BB cytoskeleton, it is no surprise that it should be among the first proteins to appear in the presumptive BB domain. As early as embryonic day 3.5, F-actin, visualized by immunofluorescence, is seen to be expressed throughout the enterocyte, although most notably in the apical domain in the cell (Noda and Mitsui, 1988; Heintzelman and Mooseker, 1990b). Villin, the multifunctional actin-binding protein of the MV core, is also expressed in day 3.5 embryonic gut, its distribution coincident with that of F-actin (Heintzelman and Mooseker, 1990b). Thus, even in the absence of a BB, or even a cell surface with many MV, actin and villin are already colocalized. From immunogold localization studies it is apparent that villin is present in the few MV that do exist, but is localized, predominantly, to the cortex of the balloon-like apex of the cell (Heintzelman and Mooseker, 1990b). In addition, prominent villin staining is seen at the level of the zonula adherans, again coinciding with actin distribution and suggesting that, at this early stage of assembly, villin is binding promiscuously to cellular arrays of F-actin.

The other major MV core proteins, fimbrin and BB myosin I, are expressed in the embryonic enterocyte at least as early as days 7–8, but are not localized to the apical domain at that time (Shibayama et al., 1987). It is not until embryonic day 10 that fimbrin, the presumed actin bundler of the MV core, assumes an apical localization. It is interesting to note that the apical expression of fimbrin correlates well with the increase in regularity and density of MV as seen in independent morphological studies (Noda, 1981; Heintzelman and Mooseker, 1990b).

The actual mechanisms for initiation of MV assembly is still unclear, but a number of possible models exist given both the morphological and immunochemical data. One model might suggest that microvilli are nucleated from the early actin–villin array by some membrane-associated protein/complex. Expression of the constituents of the as yet uncharacterized "dense tip" material seen at the apical end of MV might trigger the organization of definitive MV from the actin–villin meshwork. Experiments by Tilney and Cardell (1970) on the reassembly of the salamander BB following hydrostatic pressure-induced disassembly suggest that the dense cap material appears early and might serve as a MV nucleation site. One possible constituent of the dense cap may be Z-protein, a 55-kDa component of the myofibrillar Z-line (Ohashi and Maruyama, 1989). The expression of Z-protein in the developing enterocyte is, however, not known.

It is possible that actin and villin alone could form MV, but this begs the question of why a 3.5-day embryonic enterocyte, replete with actin and villin, does not have a BB. In this regard, the transfection experiments of Friederich *et al.* (1989, discussed below) would suggest that it is not simply the expression of villin but rather the levels of villin expression that are important. Alternatively, some trigger, such as a regulated shift in intracellular calcium levels, might variably trigger the severing, nucleating, and/or bundling activities of villin leading to a reorganization of the actin–villin network into true MV. A third possibility, and one that seems to play a role in posthatch MV elongation, is an increase in actin expression and a shift in the G-actin:F-actin ratio in favor of actin polymerization (Stidwell and Burgess, 1986). Of course, none of these possibilities are mutually exclusive.

Another model might suggest that the actin–villin array might be organized by another protein(s), including fimbrin, into definitive MV. That fimbrin may not be the primary organizer is suggested by the studies of Shibayama *et al.* (1987), who observed fimbrin expression in the cell prior to its apical localization. If fimbrin is present in the cell, why is it not associated with the actin–villin array and with the initiation of MV assembly? Again, the answer may be one of expression level, but it is just as likely that other element or cues are involved in the reorganization and assembly of true MV.

Such cellular cues may include posttranslational modifications of one or more of the cytoskeletal elements. To pursue the fimbrin story, the recent identification of fimbrin as a homolog of the cytoplasmic phosphoprotein L-plastin raises the possibility that fimbrin may also be a kinase substrate (de Arruda *et al.*, 1990). To date, phosphorylation of BB fimbrin has not been reported, nor has its phosphorylation state in the embryonic enterocyte been identified. As phosphorylation of some other actin-bundling proteins such as band 4.9 and synapsin I, inhibits actin bundle formation (Husain-Chishti *et al.*, 1988; Bahler and Greengard, 1987), one might ask if embryonic fimbrin is initially phosphorylated and then dephosphorylated coincident with the actin bundling associated with MV formation.

Related evidence that such posttranslational modifications may be important in some aspects of enterocyte differentiation come from the studies of Burgess *et al.* (1989), who examined the expression of cytoskeleton-associated phosphotyrosine-containing proteins during enterocyte differentiation along the crypt–villus axis. They discovered two major tyrosine kinase substrates whose phosphotyrosine content was much higher in the undifferentiated crypt cells compared to the differentiated cells of the villus compartment. It is apparent, then, that the phosphorylation state of any number of BB-associated cytoskeletal elements may play some role in the differentiation of the ultrastructural framework of the enterocyte.

2. The Completion of Brush Border Assembly

Between 11 and 18 days of development, little growth of MV occurs, although the rootlets are gradually elongating to their adult length of about 0.5 μm (Overton and Shoup, 1964; Chambers and Grey, 1979). After this time, three major events lead to the establishment of adult MV (Stidwell and Burgess, 1986). These include an increase in the number of MV (between 18 and 21 days), an increase in the number of actin filaments comprising the MV core (between 1 and 2 days posthatch), and a rapid elongation of the MV (days 2–4 posthatch). Stidwell and Burgess (1986) present evidence suggesting that increases in the level of cellular actin and a shift in the G-actin:F-actin ratio in favor of actin polymerization may participate in these final events seen in MV assembly.

It is also during the final days of development that the distribution of BB myosin I, the MV core–membrane linker, also changes. Prior to day 19, BB myosin I is detected throughout the embryonic epithelium and exhibits only low levels of expression in the MV domain (Shibayama et al., 1987; M. B. Heintzelman and M. S. Mooseker, unpublished observations). However, between embryonic day 18 and the time of hatch, BB myosin I exhibits a sequential redistribution first to the cell periphery and then to the apical domain (Shibayama et al., 1987). Thus, localization of BB myosin I to the BB domain (predominantly microvillar) coincides with both the final structural maturation of the MV cytoskeleton as discussed above, and also with functional maturation as assessed by the expression of membrane hydrolases whose appearance peaks at the time of hatch (Black and Moog, 1978). It appears more likely, then, that BB myosin I plays some as yet undefined role in either the structural or functional maturation of the BB rather than in the initiation of BB cytoskeletal assembly seen during the first week of development.

B. Immunochemical and Localization Studies of Terminal Web Protein Expression in the Chick Embryo

In contrast to some aspects of MV assembly, which sees the expression of the major proteins prior to their recruitment into actual MV structures, the expression of TW proteins seems to correlate with their assembly into the TW structure itself. In the chicken, three major TW components have been identified as interrootlet cross-linkers in the adult BB. These are myosin II and two distinct forms of nonerythroid spectrin: the widely distributed fodrin form and another member of the spectrin family unique to the chicken intestinal BB, TW260/240 (Hirokawa *et al.*, 1982, 1983; Glenney and Glenney, 1983; Glenney *et al.*, 1982). In the embryonic day 13 enterocyte, the existing MV generally have short rootlets and some interrootlet cross-linkers (Takemura *et al.*, 1988). Immunofluorescence and immunoblot analysis using antibodies specific for fodrin 240, TW260, and myosin II revealed that low levels of fodrin and myosin were already present in the TW region at this early stage, but the TW260/240 form was not yet expressed (Glenney and Glenney, 1983; Takemura *et al.*, 1988). Between days 13 and 19, low levels of TW260/240 were expressed and localized to the TW domain although adult-like levels of expression were not seen until approximately 2 days after hatching. During these last few days of development, an increase in the actin-associated cross-linkers was observed as final stratification of the TW domain occurred (Chambers and Grey, 1979; Takemura *et al.*, 1988).

Unlike the delayed incorporation of core proteins into the MV, TW proteins are expressed coincident with structural assembly. Given that myosin II and fodrin are present very early in the TW domain, Takemura *et al.* (1988) suggest that these proteins are involved in some fundamental aspect of rootlet cross-linking, unlike TW260/240, whose late expression occurs some time after rootlets have appeared. They suggest that TW260/240 may be a necessary stabilizing component to allow for the rapid elongation of MV that occurs soon after hatching, but, as mentioned previously, the lack of a TW as extensive as that seen in the chicken intestinal brush border does not preclude the formation of long MV in the BB domains of other cell types. As with the MV core proteins, experimentally controlled expression of the TW components and analysis of their *in vivo* effects will be necessary to accurately assess their exact roles in the BB cytoskeleton as a whole.

C. Immunochemical and Localization Studies of Brush Border Proteins in Mammalian Embryos

The comparison between the avian and the mammalian systems has thus far suggested that a common pattern of BB assembly exists. The available immunocytochemical studies on mammalian systems continue to bear this out. Studies of the early mouse embryo indicate that villin expression begins soon after the formation of the gut anlage, at day 8 of development (Maunoury *et al.*, 1988). However, apical staining in the presumptive BB domain does not occur until 12–13 days of development even though, at this time, the embryonic enterocyte lacks even the hint of a BB.

Fimbrin, the ubiquitous F-actin-bundling protein, is expressed throughout development in both embryonic and extraembryonic tissues (Ezzell *et al.*, 1989). Thus it is actually expressed in the gut prior to villin, but, as was the case in the chick embryo, localization of fimbrin to the apical domain of the enterocyte lagged behind villin by 2–3 days, just about the time when the enterocyte began to assemble MV in earnest. A similar pattern of the sequential expression of villin and fimbrin has been noted in the development of the visceral endoderm of the embryonic yolk sac (Ezzell *et al.*, 1989; Maunoury *et al.*, 1988). This is another structure that assembles an impressive array of MV for uses in embryonic nutritional exchanges prior to the establishment of the haemochorial placenta (Rugh, 1968).

The story for mammalian BB myosin I is more complicated than that of the chicken (Rochette-Egly and Haffen, 1987). A 135-kDa protein, cross-reactive with a polyclonal antibody made against avian BB myosin I, is expressed in the embryonic day 13 rat intestine even though no BB myosin I (110 kDa) is evident by immunoblot analysis. An appropriate 110-kDa band does not appear until day 19 of development in the rat, just 3 days prepartum. After birth, however, the 135-kDa form is lost and the 110-kDa BB myosin I becomes the only immunoreactive species (Rochette-Egly and Haffen, 1987). Curiously, no immunofluorescence staining of the gut is evident with the BB myosin I antibody until day 20, when the 110-kDa form is expressed, even though the 135-kDa form has been detected on immunoblots as early as day 13. However, the available immunofluorescence staining does illustrate an initial diffuse staining of BB myosin I at embryonic day 20 followed by apical localization in the adult BB. Thus, the late redistribution of BB myosin I in the rat is the same pattern observed in the avian system. A similar pattern of myosin I expression is seen in the human gut except that, in both small and large intestine, a 130-kDa immunogen persists in the adult gut (M. S. Mooseker and J. M. Carboni, unpublished observations). It is not yet clear what the significance of the 135-kDa protein is, although its cross-reactivity with BB myosin I antibodies suggests that it may be an additional form of myosin I and, in the rat, perhaps an embryonic-specific isoform of myosin I.

The story for calmodulin expression during mammalian development (both rat and human) is similar to that for the expression of the 110-kDa heavy chain of myosin I. That is, calmodulin levels are high throughout development, as assessed by immunoblotting, but its localization is not obvious until day 18 in the rat, when some weak immunofluorescence is seen in the primitive BB domain (Rochette-Egly *et al.*, 1986). Brush border-associated calmodulin staining increases after day 18, paralleling the structural differentiation of the BB. Thus, localization of calmodulin to the BB is another one of the late events in BB assembly, presumably because the bulk of CM assembled into the BB is bound to the heavy chain of BB myosin I, which itself localizes late in development.

V. Assembly of the Brush Border Cytoskeleton during Enterocyte Differentiation in the Adult Intestine

As alluded to previously, the embryonic intestine is not the only model system for studying BB assembly in the enterocyte. In the adult, the intestinal epithelium is undergoing continual renewal as stem cells located in the intestinal crypts divide and then differentiate as they migrate up the crypt–villus (C–V) axis. After 2 to 4 days, depending on the animal and the section of gut, the enterocytes are exfoliated from the villus tip (Cheng and Leblond, 1974; Lipkin, 1973). As adult cells need to differentiate much more quickly than in the embryo, one would expect that the stages of assembly seen clearly in the embryo would be greatly accelerated in the adult. Morphological and immunocytochemical studies bear out this assumption.

A. Morphological Studies

Studies from rat, human, and chicken intestines all illustrate the same basic pattern of crypt cell development (Trier, 1963; Toner, 1968; van Dongen et al., 1976; Fath et al., 1990). At the base of the crypt, the proliferative zone, one sees the beginnings of absorptive cell differentiation with the appearance of short (0.5 μm) and sparse MV at the apical surface of the cells. These early MV are oriented at angles to the cell surface and, unlike the embryonic MV, contain a loosely bundled array of filaments that extends up to 1 μm or more into the subjacent cytoplasm. Little other terminal web structure is evident in these primitive cells. However, as the cells migrate upward through the crypt and toward the villus compartment, the MV become more numerous and much more uniformly organized. During this time, stratification of the TW occurs and establishes the zone of exclusion typical of the mature enterocyte.

B. Patterns of Protein Expression along the Crypt–Villus Axis

The majority of immunochemical studies examining protein expression and distribution along the C–V axis indicate that the major BB-associated proteins (both MV and TW) are expressed in the crypt cells, and, with the exception of BB myosin I and calmodulin, are already localized to the apical domain of the developing enterocyte. In the chicken, Heintzelman and Mooseker (1990b) localized the major MV core proteins during C–V differentiation and observed that while actin, villin, and fimbrin were all localized to the apical domain in crypt cells, BB myosin I, although expressed in the crypt, did not localize to the BB domain until the C–V transition zone. Fath et al. (1990) described some MV core–membrane cross-linkers in the early crypt MV, but corroborated that BB myosin I staining was not localized to the apical domain this early.

Using an enzyme-linked immunosorbent assay (ELISA) of cell lysates isolated along the C–V axis, Fath et al. (1990) also determined levels of protein expression for actin, myosin I and myosin II, spectrin, tropomyosin, and α-actinin, but did not see any significant changes along the C–V axis. Curiously, no change in the G-actin:F-actin ratio similar to that seen during MV elongation in the chick embryo was observed along the C–V axis even though substantial growth of MV does occur here as well. Again, with the exception of BB myosin I, Fath *et al.* (1990) observed that actin, villin, myosin II, tropomyosin, and spectrin were all localized to the apical domain along the entire C–V axis of the chicken, further suggesting that BB assembly is occurring quite rapidly. Consequently, the individual assembly events seen during embryogenesis with respect to the sequential appearance of both MV and TW proteins cannot be resolved during the rapid development of crypt cells, the one exception being the late appearance of BB myosin I seen in both models of BB assembly.

Similar studies of enterocyte differentiation along the C–V axis in mammals are not as complete as in the avian system, but have demonstrated that villin is expressed apically along the entire axis (Robine *et al.*, 1985; Boller *et al.*, 1988) while calmodulin does not assume its apical localization until the villus compartment, mirroring the pattern seen for BB myosin I in the chicken gut (Rochette-Egly *et al.*, 1988). One discrepancy between the avian and mammalian system was demonstrated by Younes *et al.* (1989), who did observe a crypt–villus transition in the expression of fodrin in the human gut. Although fodrin was expressed on the basolateral domain of crypt cells, apical localization was not observed until cells entered the villus compartment. In this regard, it will be interesting to examine fodrin expression in the embryonic mammalian gut to see if its incorporation into the developing BB is as late an event as would be suggested by the findings of Younes *et al.* (1989).

C. Patterns of RNA Expression along the Crypt–Villus Axis

Brush border assembly along the C–V axis has also been examined at the level of mRNA expression. As might be expected, given the up-regulation of villin protein expression during the emergence of cells from the crypt, villin mRNA levels increase noticeably at the villus base but decline a bit further up the villus axis (Boller *et al.*, 1988). Fath *et al.* (1990) have also noted increases in transcripts for villin, calmodulin, and tropomyosin levels along the C–V axis, but saw no changes in mRNA levels for actin, the α-subunit of fodrin (the subunit shared with TW260/240) or myosin II. In an interesting series of experiments, Cheng and Bjerknes (1989) also detected little increase in the actin mRNA along the C–V axis, but did see an intracellular redistribution of actin mRNA in concert with BB differentiation along the C–V axis. That is, while actin mRNA was diffusely distributed in cells of the lower crypt, it gradually accumulated in the apical domain of the developing enterocyte. Whether the mRNA is itself interacting with actin and, if so, what the significance of this might be in terms of transcriptional regulation and thus actin dynamics is unknown (also see Ornelles *et al.*, 1986; Singer *et al.*, 1989). Hopefully, similar analyses for other BB cytoskeletal proteins will help to assess the role of transcript localization in BB cytoskeletal dynamics.

VI. *In Vitro* Models for Intestinal Brush Border Differentiation

Experimental manipulations of enterocyte differentiation with the goal of increasing our understanding of cytoskeletal assembly are necessarily limited in *in vivo* systems. Fortunately, a number of cell lines are available and have already been used extensively for a variety of biochemical and physiological studies (e.g., Laburthe *et al.*, 1980; Hidalgo *et al.*, 1989; Madara and Stafford, 1989). Although a number of these cell lines can assemble an apical BB domain, the Caco2 and the HT-29 cell lines have thus far been the only ones used as *in vitro* models for studying the organization and assembly of the BB cytoskeleton during cell differentiation (Pinto *et al.*, 1982, 1983; Le Bivic *et al.*, 1988).

The HT-29 cell line provides a useful system in which to study BB assembly, as this cell line can undergo a reversible differentiation in response to specific manipulations of the culture medium. That is, the removal of glucose from the medium will trigger differentiation of the cell, including both the assembly of an apical brush border and the expression of apical membrane hydrolases (Pinto *et al.*, 1982). Using this model of regulated differentiation, it has been shown that the expression of both villin mRNA and its product increased coincident with differentiation, and that the protein assumed an apical distribution unlike the diffuse distribution seen in the undifferentiated cells (Prignault *et al.*, 1986; Robine *et al.*, 1985; Dudouet *et al.*, 1987). The potential for this cell line in the study of other aspects of BB assembly, disassembly, and maintenance are obvious, and future studies should yield some important insights into assembly mechanisms.

A second model that has been used for studies of BB assembly in the Caco2 cell line, which, under standard culture conditions, will assemble a BB domain (Pinto *et al.*, 1983; Hidalgo *et al.*, 1989). The BB cytoskeleton of the differentiated Caco2 cell line has been shown by both immunochemical and immunolocalization techniques to contain villin, fimbrin, both 110-kDa and 130-kDa myosin I forms, fodrin, and myosin II (Peterson and Mooseker, 1990; Robine *et al.*, 1985). Furthermore, BB assembly during Caco2 cell differentiation appears to resemble, at least morphologically, that described for intestinal BB development *in vivo*. Immunocytochemically, a pre-MV cortical array of actin and villin is seen in the subconfluent Caco2 cells (Peterson and Mooseker, 1990, and unpublished observations) reminiscent of that seen during *in vivo* BB formation in the embryonic chick gut (Heintzelman and Mooseker, 1990b).

A third *in vitro* model that has been utilized to dissect possible mechanisms of BB assembly is that described by Friederich *et al.* (1989). With a goal toward understanding what role villin might play in the initiation of BB assembly, Friederich *et al.* (1989) transfected a fibroblast cell line with different constructs of a cDNA clone for human villin. They illustrated that low levels of villin expression resulted in the colocalization of villin with the cortical actin cytoskeleton. However, increased levels of villin expression dramatically changed the morphology of the apical domain as numerous, long, stiff MV appeared concomitant with the disappearance of stress fibers. It is still uncertain whether villin itself triggered the assembly of new MV or whether it simply stabilized and enhanced the growth of preexisting MV. It will also be important to assess the role of other actin-binding proteins in the villin-induced rearrangements of the actin-based cytoskeleton. For example, are other bundling proteins (e.g., fimbrin) recruited into the new MV? It is clear that the use of this transfection model in looking at the roles, both separately and in combinations, of the other BB cytoskeletal proteins should continue to yield some important insights into how the cell orchestrates the assembly of this cytoskeletal array.

VII. Mechanisms of Maintenance and Reassembly of the Brush Border Cytoskeleton

Once the intestinal BB cytoskeleton is assembled after embryonic development or C–V axis differentiation, it does not survive simply as a static structure for the maintenance of the apical domain. Rather, the BB cytoskeleton is a dynamic array of proteins that undergoes a constant turnover. Metabolic labeling of BB proteins *in vivo* was accomplished by Stidwell *et al.* (1984), who pulsed chickens with [^{35}S]methionine. High levels of BB labeling were seen at the first 6-hr timepoint, with lower levels of protein remaining at subsequent 6-hr timepoints. They observed similar turnover rates for actin, villin, fimbrin, and myosin (II) while BB myosin I heavy chain (110 kDa) and TW260/240 had somewhat faster turnover rates. Compared to the turnover rates of several hours for the cytoskeletal proteins, the BB membrane proteins of mammals are known to turn over in less than an hour (Hauri *et al.*, 1979; Quaroni *et al.*, 1979).

That synthesis and turnover rates of the cytoskeletal vs membrane protein populations are quite different has also been noted by Cowell and Danielson (1984). They metabolically labeled BB cytoskeletal proteins in explants of pig intestine and observed that labeled cytoskeletal proteins appeared in the BB almost immediately (10 min) while membrane proteins showed a delayed appearance of 40–60 min, consistent with their requisite transport through the endoplasmic reticulum (ER) and Golgi. In any case, it is remarkable that the enterocyte can maintain such a uniform array as the BB in spite of the continual renewal of both its cytoskeletal and membrane constituents.

The structural dynamism of the BB cytoskeleton also becomes obvious when one looks at the responses of the BB to physiological cues such as feeding and fasting. Misch *et al.* (1980) describes the tall MV and MV-derived vesicular elements seen in the adlumenal surface of the gut of well-fed animals. In contrast, the MV of fasted animals were 25–30% shorter. Similar MV shortening was observed in rats challenged with the lectins of raw kidney beans (Weinmann *et al.*, 1989). Two to 4 hr following administration of a kidney bean diet, MV were seen to vesiculate and, by 6 hr, had become uniformly shorter and fatter than the controls. When allowed to recover, the MV reassumed their pre-challenge morphology within 20 hr. Similar results have been achieved *in vitro* in the Caco2 cell line challenged with lectins (Draaijer *et al.*, 1989). In addition to observing the shortening and fattening of MV, Draaijer *et al.* (1989) also observed an increase in the percentage of G-actin in the cells. Characterization of the intracellular triggers for these lectin affects may establish this system as a very useful model for the study of BB cytoskeletal dynamics.

Another experimental approach for looking at the dynamics of BB assembly and disassembly was the experiment of Tilney and Cardell (1970). Subjecting an explant of salamander gut to hydrostatic pressure resulted in the loss of BB architecture as MV were lost by both retraction and fragmentation. The TW and zone of exclusion were also lost. In addition, the apical domain bulged out into the lumen of the gut, establishing a morphology reminiscent of an early embryonic chick enterocyte. On release of pressure, reassembly was observed commencing with the appearance of dense plaques on the apical membrane. Tilney and Cardell (1970) likened these to the dense cap material seen at the tips of mature microvilli and suggested that they might serve as nucleation sites for the actin filaments that were subsequently observed. Within 15 min after decompression, the number of MV increased and the TW started to reform such that by 90 min the cells appeared quite normal.

Given what we now know about the composition of the BB domain and the pattern of expression of both proteins and message levels, it would be of some interest to repeat these morphological studies with more advanced techniques to look at the reassembly process at the molecular level. Unfortunately, repeated attempts at duplicating these experiments in the well-defined avian system have been unsuccessful (M. Mooseker, D. Begg, and C. Howe, unpublished observations).

VIII. The Intestinal Brush Border Cytoskeleton in Disease

In addition to the characterization of the brush border cytoskeleton in cultured colonic adenocarcinoma cell lines, some recent work has also focused on the BB cytoskeleton, particularly fodrin and villin, in resection or biopsy samples of human colonic adenocarcinomas. Younes *et al.* (1989) have used digital confocal microscopy to examine fodrin expression and distribution in both adenomas and adenocarcinomas of the colon. They observed not only a threefold increase in total cellular fodrin, but also enhanced staining of fodrin in the BB domain of affected cells. They suggest that increased apical staining of fodrin may be related to the ultrastructural changes occurring in the BB such as MV shortening and enhancement of the MV rootlet domain (Imai and Stein, 1963; Cardoso *et al.*, 1971; Younes *et al.*, 1989). Increases in total fodrin expression might simply reflect neoplastic activity because this phenomenon was observed in a variety of epithelial neoplasms (Younes *et al.*, 1989).

In contrast to the widespread distribution of fodrin, the limited distribution of villin in the normal organism (Robine *et al.*, 1985) raises the hope that this protein might be used not only as a marker for determining the primary origin of some metastatic growths but also the degree of differentiation of some tumors (e.g., Moll *et al.*, 1987). A number of groups have looked at villin expression in human colonic adenocarcinomas and, with the possible exception of the most undifferentiated tumors, villin is still expressed and maintains its BB localization (Robine *et al.*, 1985; West *et al.*, 1988; Carboni *et al.*, 1987; Moll *et al.*, 1987; Grone *et al.*, 1986). Some subtle changes, such as elevation of villin expression in the cytoplasm and along the basement membrane, have been noted (West *et al.*, 1988; Carboni *et al.*, 1987) but these have not been consistent findings as villin immunoreactivity has varied between tumors even of the same grade (Carboni *et al.*, 1987). In addition, villin expression in primary adenocarcinoma of the endometrium and the lung has been observed even though villin is not usually expressed in these tissues (Moll *et al.*, 1987). Clearly, further characterization of villin expression in cancerous tissue, perhaps more closely correlated with the ultrastructural changes associated with neoplastic changes, will be necessary before this marker can be used in an effective way for immunocytochemical studies. However, recent work by Dudouet *et al.* (1990) would suggest that villin may still be a useful tool for both the diagnosing and monitoring of patients with colorectal carcinomas, because a strong correlation does exist between these conditions and increased levels of villin in the sera of these patients.

Another aspect of BB biology that appears to have a pathological correlate is the aberrant BB assembly seen in congenital microvillus atrophy, also known as microvillus inclusion disease (MID; Davidson *et al.*, 1978). Presenting as an intractable diarrhea in infancy, MID is characterized, histologically, by villus atrophy, hypoproliferation of the crypt cells, and aberrant assembly of the BB cytoskeleton in absorptive cells of the villus. In addition to the sparse, irregular MV on the apical domain of the domed enterocytes, the cells are replete with cytoplasmic inclusions, largely of lysosomal origin, but including the distinctive cysts complete with an intact BB cytoskeleton (Davidson *et al.*, 1978; Carruthers *et al.*, 1985; Cutz *et al.*, 1989). In contrast, the cytoskeleton of the crypt cells and the morphology of the other intestinal cell types (goblet, Paneth, and enteroendocrine) appear to be well preserved (Davidson *et al.*, 1978). The cause of this disease is uncertain, but it has been suggested that MID represents a defect in proper BB assembly and differentiation related to an inborn error of intracellular transport (Cutz *et al.*, 1989).

Support for such a theory comes from an *in vivo* experimental model that illustrates the effects of drug-induced microtubule disruption on the assembly of the intestinal BB cytoskeleton (Pavelka *et al.*, 1983; Buschmann, 1983; Achler *et al.*, 1989). When rats and mice were treated with colchicine or vinblastine, BB cytoskeletal elements appeared on the basolateral membrane within 6 hr. By 12 hr, the misplaced BBs had formed intracellular inclusions that, over the course of the next 24 hr, migrated to the apical domain to finally fuse with each other and with the apical membrane. Immunocytochemical studies suggested that the major MV core proteins actin, villin, fimbrin, and BB myosin I, and a number of BB membrane enzymes, were present in both the basolateral BBs and in the BB inclusions. Thus, the BB cytoskeleton seems to be assembling properly, but does so on the wrong membrane domain (Achler *et al.*, 1989). The similarities between MID and the experimental model are intriguing and suggest that the "defective intracellular transport" theory proposed for MID should be investigated in more detail.

IX. Prospects for the Future

It is obvious from the preceding pages that much is known about the organization, composition, and assembly of the brush border cytoskeleton. However, speculative hypotheses rather than mechanistic descriptions still define our current state of knowledge regarding how the enterocyte actually assembles and maintains the BB cytoskeleton. With a few exceptions, the studies reviewed here have been largely descriptive and, although they give us some vital information about when, where, and how much a given protein is expressed during the assembly process, we are still left on our own to synthesize the best, current model for assembly.

All of the systems described thus far still have more information to offer us, but for the immediate future, analysis of BB structure, function, and assembly at the molecular level is necessary to fully understand the *in vivo* roles and interactions among the major, and hopefully minor, BB constituents. For this reason, continued use of the transfection models introduced by Friederich *et al.* (1989) and similar studies using the established BB expressing cell lines should allow us to manipulate and evaluate the individual and cooperative roles of the BB proteins. The major players have all been well characterized *in vitro,* and we understand the range of potential interactions among them, but *in vivo* studies are the necessary next step in understanding what actually occurs rather than what might occur.

In this regard, another system with a great deal of potential is the BB found in the larval midgut of *Drosophila* and other anthropods. The midgut of *Drosophila* larvae contains an impressive BB (e.g., Dimitriadis and Kastritsis, 1984), and preliminary evidence suggests that villin, myosin, and spectrin are present as BB cytoskeletal components (M. B. Heintzelman, M. S. Mooseker, D. Louvard, and S. Artavanis-Tsakonas, unpublished observations). In fact, recent work on the larval midgut of *Manduca* has revealed that the structural and biochemical

116 Matthew B. Heintzelman and Mark S. Mooseker

composition of the insect BB is quite similar to that already described for vertebrate systems (Camatini et al., 1990). Obviously, cloning of the major BB genes
in *Drosophila* and analysis of mutations, in particular deletions, will finally
allow us to discern what is necessary and what is sufficient for the assembly and
maintenance of the BB cytoskeleton.

Acknowledgments

We wish to thank our colleagues in the brush border field who sent us reprints and preprints of their
work. Thanks also to the members of the Mooseker laboratory for their help and support throughout
this project. Special thanks to Kris Mooseker and Dr. Rick Fehon for their help with the illustrations.

References

Achler, C., Filmer, D., Merte, C., and Drenckhahn, D. (1989). Role of microtubules in polarized
delivery of apical membrane proteins to the brush border of the intestinal epithelium. *J. Cell
Biol.* **109**, 179–189.
Arikawa, K., Hicks, J. L., and Williams, D. S. (1990). Identification of actin filaments in the
rhabdomeral microvilli of *Drosophila* photoreceptors. *J. Cell Biol.* **110**, 1993–1998.
Bahler, M., and Greengard, P. (1987). Synapsin I bundles F-actin in a phosphorylation-dependent
manner. *Nature (London)* **326**, 704–707.
Bailey, D. S., Cook, A., McAllister, G., Moss, M., and Mian, N. (1984). Structural and
biochemical differentiation of the mammalian small intestine during fetal development. *J. Cell
Sci.* **72**, 195–212.
Bell, L., and Williams, L. (1982). A scanning and transmission electron microscopical study of
the morphogenesis of human colonic villi. *Anat. Embryol.* **165**, 437–455.
Black, B. L., and Moog, F. (1978). Alkaline phosphatase and maltase activity in the embryonic
chick intestine in culture. Influences of thyroxine and hydrocortisone. *Dev. Biol.* **66**, 232–249.
Blest, A. D., Stowe, S., and Eddey, W. (1982). A labile, Ca^{++}-dependent cytoskeleton in
rhabdomeral microvilli of blowflies. *Cell Tissue Res.* **223**, 553–573.
Boller, K., Arpin, M., Pringault, E., Mangeat, P., and Reggio, H. (1988). Differential
distribution of villin and villin mRNA in mouse intestinal epithelial cells. *Differentiation
(Berlin)* **39**, 51–57.
Bretscher, A. (1983). Purification of an 80,000-dalton protein that is a component of the isolated
microvillus cytoskeleton, and its localization in nonmuscle cells. *J. Cell Biol.* **97**, 425–432.
Bretscher, A. and Lynch, W. (1985). Identification and localization of immunoreactive forms of
caldesmon in smooth and nonmuscle cells: A comparison with the distributions of tropomyosin
and α-actinin. *J. Cell Biol.* **100**, 1656–1663.
Bretscher, A., and Weber, K. (1978). Localization of actin and microfilament-associated proteins
in the microvilli and terminal web of the intestinal brush border by immunofluorescence
microscopy. *J. Cell Biol.* **79**, 839–845.
Bretscher, A., and Weber, K. (1980). Fimbrin, a new microfilament-associated protein present in
microvilli and other cell surface structures. *J. Cell Biol.* **86**, 335–340.
Burgess, D. R. (1975). Morphogenesis of intestinal villi. II. Mechanism of formation of
previllous ridges. *J. Embryol. Exp. Morphol.* **34**, 723–740.
Burgess, D. R. (1982). Reactivation of intestinal epithelial cell brush border motility: ATP-
dependent contraction via a terminal web contractile ring. *J. Cell Biol.* **95**, 853–863.

Burgess, D. R., Broschat, K. O., and Hayden, J. M. (1987). Tropomyosin distinguishes between the two actin-binding sites of villin and affects actin-binding properties of other brush border proteins. *J. Cell Biol.* **104**, 29–40.

Burgess, D. R., Jiang, W., Mamajiwalla, S., and Kinsey, W. (1989). Intestinal crypt stem cells possess high levels of cytoskeletal-associated phosphotyrosine-containing proteins and tryosine kinase activity relative to differentiated enterocytes. *J. Cell Biol.* **109**, 2139–2144.

Buschmann, R. J. (1983). Morphometry of the small intestinal enterocytes of the fasted rat and the effects of colchicine. *Cell Tissue Res.* **231**, 289–299.

Camatini, M., Maci, R., Giordana, B., Khawaja, S., Colkombo, A., and Cappelletti, G. (1990). Cytoskeletal elements of the brush border from the midgut to lepidopteran larvae: Preliminary results. *J. Cell Biol.* **109**, 172a.

Carboni, J. M., Howe, C. L., West, A. B., Barwick, K. W., Mooseker, M. S., and Morrow, J. S. (1987). Characterization of intestinal brush border cytoskeletal proteins of normal and neoplastic human epithelial cells. A comparison with the avian brush border. *Am. J. Pathol.* **129**, 589–600.

Cardoso, J. M. J., Diener, K. A., Alvarez, E. E., and Maldonado, E. M. (1971). Electron microscopy of adenocarcinoma of the colon. *Am. J. Proct.* **22**, 301–307.

Carruthers, L., Phillips, A. D., Dourmashkin, R., and Walker-Smith, J. A. (1985). Biochemical abnormality in brush border membrane protein of a patient with congenital microvillus atrophy. *J. Pediatr. Gastroenterol. Nutr.* **4**, 902–907.

Chambers, C., and Grey, R. D. (1979). Development of structural components of the brush border in absorptive cells of the chick intestine. *Cell Tissue Res.* **204**, 387–405.

Cheng, H., and Bjerknes, M. (1989). Asymmetric distribution of actin mRNA and cytoskeletal pattern generation in polarized epithelial cells. *J. Mol. Biol.* **210**, 541–549.

Cheng, H., and Leblond, C. P. (1974). Origin, differentiation and renewal of the four main epithelial cell types in the mouse small intestine. I. Columnar cell. *Am. J. Anat.* **141**, 461–480.

Collins, K., Sellers, J. R., and Matsudaira, P. (1990). Calmodulin dissociation regulates brush border myosin I (110-kD-calmodulin) mechanochemical activity *in vitro*. *J. Cell Biol.* **110**, 1137–1147.

Coluccio, L. M., and Bretscher, A. (1989). Reassociation of microvillar core proteins: Making a microvillar core *in vitro*. *J. Cell Biol.* **108**, 495–502.

Coudrier, E., Kerjaschki, D., and Louvard, D. (1988). Cytoskeletal organization and submembranous interactions in intestinal and renal brush borders. *Kidney Int.* **34**, 309–320.

Coulombre, A. J., and Coulombre, J. L. (1958). Intestinal development. I. Morphogenesis of the villi and musculature. *J. Embryol. Exp. Morphol.* **6**, 403–411.

Cowell, G. M., and Danielsen, E. M. (1984). Biosynthesis of intestinal microvillar proteins. Rapid expression of cytoskeletal components in microvilli of pig small intestine mucosal explants. *FEBS Lett.* **172**, 309–314.

Cutz, E., Rhoads, J. M., Drumm, B., Sherman, P. M., Durie, P. R., and Forstner, G. G. (1989). Microvillus inclusion disease: An inherited defect of brush border assembly and differentiation. *N. Engl. J. Med.* **320**, 646–651.

Davidson, G. P., Cutz, E., Hamilton, J. R., and Gall, D. G. (1978). Familial enteropathy: A syndrome of protracted diarrhea from birth, failure to thrive, and hypoplastic villus atrophy. *Gastroenterology* **75**, 783–790.

de Arruda, M. V., Watson, S., Lin, C.-S., Leavitt, J., and Matsudaira, P. (1990). Fimbrin is a homologue of the cytoplasmic phosphoprotein plastin and has domains homologous with calmodulin and actin gelation proteins. *J. Cell Biol.* **111**, 1069–1079.

Dimitriadis, V. K., and Kastritsis, C. D. (1984). Ultrastructural analysis of the midgut of *Drosophila auraria* larvae. Morphological observations and their physiological implications. *Can. J. Zool.* **62**, 659–669.

Draaijer, M., Koninkx, J., Hendriks, H., Kik, M., van Dijk, J., and Mouwen, J. (1989). Actin cytoskeleton lesions in differentiated human colon carcinoma Caco-2 cells after exposure to soybean agglutinin. *Biol. Cell.* **65,** 29–35.

Drenckhahn, D., and Dermietzel, R. (1988). Organization of the actin filament cytoskeleton in the intestinal brush border: A quantitative and qualitative immunoelectron microscope study. *J. Cell Biol.* **107,** 1037–1048.

Drenckhahn, D., and Gröschel-Stewart, U. (1980). Localization of myosin, actin, and tropomyosin in rat intestinal epithelium: immunohistochemical studies at the light and electron microscope levels. *J. Cell Biol.* **86,** 475–482.

Dudouet, B., Robine, S., Huet, C., Sahuquillo-Merino, C., Blair, L., Coudrier, E., and Louvard, D. (1987). Changes in villin synthesis and subcellular distribution during intestinal differentiation of HT29-18 clones. *J. Cell Biol.* **105,** 359–369.

Dudouet, B., Jacob, L., Beuzeboc, P., Magdelenat, H., Robine, S., Chapuis, Y., Christoforov, B., Cremer, G. A., Pouillard, P., Bonnichon, P., Pinon, F., Salmon, R. J., Pointereau-Bellanger, A., Bellanger, J. Maunoury, M. T., and Louvard, D. (1990). Presence of villin, a tissue-specific cytoskeletal protein, in sera of patients and an initial clinical evaluation of its value for the diagnosis and follow-up of colorectal cancers. *Cancer Res.* **50,** 438–443.

Ezzell, R. M., Chafel, M. M., and Matsudaira, P. T. (1989). Differential localization of villin and fimbrin during development of the mouse visceral endoderm and intestinal epithelium. *Development (Cambridge, UK)* **106,** 407–419.

Fath, K. R., Obenauf, S. D., and Burgess, D. R. (1990). Cytoskeletal protein and mRNA accumulation during brush border formation in adult chicken enterocytes. *Development (Cambridge, UK)* **109,** 449–459.

Friederich, E., Huet, C., Arpin, M., and Louvard, D. (1989). Villin induces microvilli growth and actin redistribution in transfected fibroblasts. *Cell (Cambridge, Mass.)* **59,** 461–475.

Friederich, E., Pringault, E., Arpin, M., and Louvard, D. (1990). From the structure to the function of villin, an actin-binding protein of the brush border. *Bioessays* **12,** 403–408.

Gerke, V., and Weber, K. (1984). Identity of p36K phosphorylated upon Rous sarcoma virus transformation with a protein purified from brush borders: Calcium-dependent binding to nonerythroid spectrin and F-actin. *EMBO J.* **3,** 227–233.

Glenney, J. R., and Glenney, P. (1983). Fodrin is the general spectrin-like protein found in most cells whereas spectrin and the TW protein have a restricted distribution. *Cell (Cambridge, Mass.)* **34,** 503–512.

Glenney, J. R., Kaulfus, P., Matsudaira, P., and Weber, K. (1981). F-actin binding and bundling properties of fimbrin, a major cytoskeletal protein of the microvillus core filaments. *J. Biol. Chem.* **256,** 9283–9288.

Glenney, J. R., Glenney, P., Osborn, M., and Weber, K. (1982). An F-actin- and calmodulin-binding protein from isolated intestinal brush borders has a morphology related to spectrin. *Cell (Cambridge, Mass.)* **28,** 843–854.

Gould, K. L., Cooper, J. A., Bretscher, A., and Hunter, T. (1986). The protein kinase substrate, p81, is homologous to a chicken microvillar core protein. *J. Cell Biol.* **102,** 660–669.

Greenberg, M. E., Brackenbury, R., and Edelman, G. M. (1984). Changes in the distribution of the 34-kdalton tryosine kinase substrate during differentiation and maturation of chicken tissues. *J. Cell Biol.* **98,** 473–486.

Grey, R. D. (1972). Morphogenesis of intestinal villi. I. Scanning electron microscopy of the duodenal epithelium of the developing chick embryo. *J. Morphol.* **137,** 193–214.

Grone, H. J., Weber, K., Helmchen, U., and Osborn, M. (1986). Villin- a marker of brush border differentiation and cellular origin in human renal cell carcinoma. *Am. J. Pathol.* **124,** 294–302.

Hagen, S. J., Allan, C. H., and Trier, J. S. (1987). Demonstration of microtubules in the terminal web of mature absorptive cells from the small intestine of the rat. *Cell Tissue Res.* **248,** 709–712.

Hauri, H.-P., Quaroni, A., and Isselbacher, K. J. (1979). Biogenesis of intestinal plasma membrane: Post-translational route and cleavage of sucrose-isomaltase. *Proc. Natl. Acad. Sci. U.S.A.* **76**, 5183–5186.

Heintzelman, M. B., and Mooseker, M. S. (1990a). Assembly of the brush border cytoskeleton: Changes in the distribution of microvillar core proteins during enterocyte differentiation in adult chicken intestine. *Cell Motil. Cytoskel.* **15**, 12–22.

Heintzelman, M. B., and Mooseker, M. S. (1990b). Structural and compositional analysis of early stages in microvillus assembly in the enterocyte of the chick embryo. *Differentiation (Berlin)* **43**, 175–182.

Hidalgo, I. J., Raub, T. J., and Borchardt, R. T. (1989). Characterization of the human colon carcinoma cell line (Caco-2) as a model system for intestinal epithelial permeability. *Gastroenterology* **96**, 736–749.

Hilton, W. A. (1902). The morphology and development of intestinal folds and villi in vertebrates. *Am. J. Anat.* **1**, 459–504.

Hirokawa, N., and Heuser, J. E. (1981). Quick-freeze, deep-etch visualization of the cytoskeleton beneath surface differentiations of intestinal epithelial cells. *J. Cell Biol.* **91**, 399–409.

Hirokawa, N., Tilney, L. G., Fujiwara, K., and Heuser, J. E. (1982). The organization of actin, myosin, and intermediate filaments in the brush border of intestinal epithelial cells. *J. Cell Biol.* **94**, 425–443.

Hirokawa, N., Cheney, R. E., and Willard, M. (1983). Location of a protein of the fodrin-spectrin-TW260/240 family in the mouse intestinal brush border. *Cell (Cambridge, Mass.)* **32**, 953–965.

Husain-Chishti, A., Levin, A., and Branton, D. (1988). Abolition of actin bundling by phosphorylation of human erythrocyte protein 4.9. *Nature (London)* **334**, 718–721.

Imai, H., and Stein, A. A. (1963). Ultrastructure of the adenocarcinoma of the colon. *Gastroenterology* **44**, 410–418.

Keller, T. C. S., and Mooseker, M. S. (1982). Ca^{++}-calmodulin-dependent phosphorylation of myosin, and its role in brush border contraction *in vivo*. *J. Cell Biol.* **95**, 943–959.

Kelley, R. O. (1973). An ultrastructural and cytochemical study of developing small intestine in man. *J. Embryol. Exp. Morphol.* **29**, 411–430.

Laburthe, M., Rousset, M., Chevalier, G., Boissard, C., Dupont, C., Zweibaum, A., and Rosselin, G. (1980). Vasoactive intestinal peptide control of adenosine 3':5'-monophosphate levels in seven human colorectal adenocarcinoma cell lines in culture. *Cancer Res.* **40**, 2529–2533.

Lacroix, B., Kedinger, M., Simon-Assmann, P., Rousset, M., Zweibaum, A., and Haffen, H. (1984). Developmental pattern of brush border enzymes in the human fetal colon. Correlation with some morphogenetic events. *Early Hum. Dev.* **9**, 95–103.

Le Bivic, A., Hirn, M., and Reggio, H. (1988). HT-29 cells are an *in vitro* model for the generation of cell polarity in epithelia during embryonic differentiation. *Proc. Natl. Acad. Sci. U.S.A.* **85**, 136–140.

Lipkin, M. (1973). Proliferation and differentiation of gastrointestinal cells. *Physiol. Rev.* **53**, 891–915.

Louvard, D. (1989). The function of the major cytoskeletal components of the brush border. *Curr. Opin. Cell Biol.* **1**, 51–57.

Madara, J. L. (1987). Intestinal absorptive cell tight junctions are linked to cytoskeleton. *Am. J. Physiol.* **253**, C171–C175.

Madara, J. L. (1989). Loosening tight junctions. Lessons from the intestine. *J. Clin. Invest.* **83**, 1089–1094.

Madara, J. L., and Stafford, J. (1989). Interferon-γ directly affects barrier function of cultured intestinal epithelial monolayers. *J. Clin. Invest.* **83**, 724–727.

Mathan, M., Moxey, P. C., and Trier, J. S. (1976). Morphogenesis of fetal rat duodenal villi. *Am. J. Anat.* **146**, 73–92.

Matsudaira, P. T., and Burgess, D. R. (1979). Identification and organization of the components in the isolated microvillus cytoskeleton. *J. Cell Biol.* **83**, 667–673.

Maunoury, R., Robine, S., Prignault, E., Huet, C., Guenet, J. L., Gaillard, J. L., and Louvard, D. (1988). Villin expression in the visceral endoderm and in the gut anlage during early mouse embryogenesis. *EMBO J.* **7,** 3321–3329.

Maunsbach, A. B. (1973). Ultrastructure of the proximal tubule. *In* "Handbook of Physiology" (J. Orloff and R. W. Berlinger, eds.), Sec. 8, Vol. 31, Chapter 2, pp. 31–79. Am. Physiol. Soc., Washington, D.C.

Miller, D., and Crane, R. K. (1961). The digestive function of the epithelium of the small intestine. II. Localisation of disaccharide hydrolysis in the isolated brush border of intestinal epithelial cells. *Biochim. Biophys. Acta* **52,** 293–298.

Millington, P. F., and Finean, J. B. (1962). Electron microscope studies of the structure of the microvilli on principal epithelial cells of rat jejunum after treatment in hypo- and hypertonic saline. *J. Cell Biol.* **14,** 125–139.

Misch, D. W., Giebel, P. E., and Faust, R. G. (1980). Intestinal microvilli: Responses to feeding and fasting. *Eur. J. Cell Biol.* **21,** 269–279.

Moll, R., Robine, S., Dudouet, B., and Louvard, D. (1987). Villin: A cytoskeletal protein and a differentiation marker expressed in some human adenocarcinomas. *Virchows Arch. B* **54,** 155–169.

Mooseker, M. S. (1976). Brush border motility. Microvillar contraction in Triton-treated brush border from intestinal epithelium. *J. Cell Biol.* **71,** 417–433.

Mooseker, M. S. (1985). Organization, chemistry, and assembly of the cytoskeletal apparatus of the intestinal brush border. *Annu. Rev. Cell Biol.* **1,** 209–241.

Mooseker, M. S., and Coleman, T. R. (1989). The 110kD-protein–calmodulin complex of the intestinal microvillus (brush border myosin I) is a mechanoenzyme. *J. Cell Biol.* **108,** 2395–2400.

Mooseker, M. S., and Tilney, L. G. (1975). Organization of an actin filament–membrane complex. Filament polarity and membrane attachment in the microvilli of intestinal epithelial cells. *J. Cell Biol.* **67,** 725–743.

Mooseker, M. S., Bonder, E. M., Conzelman, K. A., Fishkind, D. J., Howe, C. L., and Keller, T. C. S. (1984). The brush border cytoskeleton and integration of cellular functions. *J. Cell Biol.* **99,** 104s–112s.

Mooseker, M. S., Conzelman, K. A., Coleman, T. R., Heuser, J. E., and Sheetz, M. P. (1989). Characterization of intestinal microvillar membrane discs: Detergent-resistant membrane sheets enriched in associated brush border myosin I (110K-calmodulin). *J. Cell Biol.* **109,** 1153–1161.

Moxey, P. C., and Trier, J. S. (1979). Development of villus absorptive cells in the human fetal small intestine: A morphological and morphometric study. *Anat. Rec.* **195,** 463–482.

Noda, S. (1981). Role of "long microvillous-like processes" in the formation of the previllous ridges of embryonic chick duodenum. *J. Tokyo Women's Med. Coll.* **51,** 615–625.

Noda, S., and Mitsui, T. (1988). Distribution of actin, vinculin, and fibronectin in the duodenum of developing chick embryos: Immunohistochemical studies at the light microscopic level. *Dev., Growth Differ.* **30,** 271–282.

Ohashi, K., and Maruyama, K. (1989). Z-protein, a component of the skeletal muscle Z-line, is located at the apical tips of microvilli of chicken intestinal epithelial cells. *J. Biochem. (Tokyo)* **106,** 115–118.

Ornelles, D. A., Fey, E. G., and Penman, S. (1986). Cytochalasin releases mRNA from the cytoskeleton and inhibits protein synthesis. *Mol. Cell Biol.* **6,** 1650–1662.

Overton, J. (1965). Fine structure of the free cell surface in developing mouse intestinal mucosa. *J. Exp. Zool.* **159,** 105–202.

Overton, J., and Shoup, J. (1964). Fine structure of cell surface specializations in the maturing duodenal mucosa of the chick. *J. Cell Biol.* **21,** 75–85.

Pavelka, M., Ellinger, A., and Gangl, A. (1983). Effect of colchicine on rat small intestinal absorptive cells. I. Formation of basolateral microvillus borders. *J. Ultrastruct. Res.* **85,** 249–259.

Pearl, M., Fishkind, D., Mooseker, M. S., Keene, D., and Keller, T. C. S. (1984). Studies on the spectrin-like protein from the intestinal brush border, TW260/240, and characterization of its interaction with the cytoskeleton and actin. *J. Cell Biol.* **98**, 66–78.

Peterson, M. D., and Mooseker, M. S. (1990). Characterization of Caco-2 cells as a model for brush border cytoskeletal assembly. *J. Cell Biol.* **111**, 165a.

Pinto, M., Appay, M-D., Simon-Assmann, P., Chevalier, G., Dracopoli, N., Fogh, J., and Zweibaum, A. (1982). Enterocytic differentiation of cultured human colon cancer cells by replacement of glucose by galactose in the medium. *Biol. Cell.* **44**, 193–196.

Pinto, M., Robine-Leon, S., Appay, M-D., Kedinger, M., Triadou, N., Dussaulx, E., Lacroix, B., Simon-Assmann, P., Haffen, K., Fogh, J., and Zweibaum, A. (1983). Enterocyte-like differentiation and polarization of the human colon carcinoma cell line Caco-2 in culture. *Biol. Cell.* **47**, 323–330.

Pollard, T. D., Doberstein, S. K., and Zot, H. G. (1991). Myosin-I. *Annu. Rev. Physiol.* **53**, 653–681.

Pringault, E., Arpin, M., Garcia, A., Finidori, J., and Louvard, D. (1986). A human villin cDNA clone to investigate the differentiation of intestinal and kidney cells *in vivo* and in culture. *EMBO J.* **5**, 3119–3124.

Quaroni, A., Kirsch, K., and Weiser, M. M. (1979). Synthesis of membrane glycoproteins in rat small-intestine cells. *Biochem. J.* **182**, 203–212.

Robine, S., Huet, C., Moll, R., Sahuquillo-Merino, C., Coudrier, E., Zweibaum, A., and Louvard, D. (1985). Can villin be used to identify malignant and undifferentiated normal digestive epithelial cells? *Proc. Natl. Acad. Sci. U.S.A.* **82**, 8488–8492.

Rochette-Egly, C., and Haffen, K. (1987). Developmental pattern of calmodulin-binding proteins in rat jejunal epithelial cells. *Differentiation (Berlin)* **35**, 219–227.

Rochette-Egly, C., Garaud, L. Cl., Kedinger, M., and Haffen, K. (1986). Calmodulin in epithelial intestinal cells during rat development. *Experientia* **42**, 1043–1046.

Rochette-Egly, C., Lacroix, B., Haffen, K., and Kedinger, M. (1988). Expression of brush border calmodulin-binding proteins during human small and large bowel differentiation. *Cell Differ.* **24**, 119–132.

Rodewald, R., Newman, S. B., and Kranovsky, M. J. (1976). Contraction of isolated brush borders from the intestinal epithelium. *J. Cell Biol.* **70**, 541–554.

Rodman, J. S., Mooseker, M. S., and Farquhar, M. G. (1986). Cytoskeletal proteins of the rat kidney proximal tubule brush border. *Eur. J. Cell Biol.* **42**, 319–327.

Rugh, R. (1968). "The Mouse, Its Reproduction and Development." Burgess, Minneapolis, Minnesota.

Sandoz, D., and Laine, M.-C. (1985). Distribution of microtubules within the intestinal terminal web as revealed by quick-freezing and cryosubstitution. *Eur. J. Cell Biol.* **39**, 481–484.

Shibayama, T., Carboni, J. M., and Mooseker, M. S. (1987). Assembly of the intestinal brush border: Appearance and redistribution of microvillar core proteins in developing chick enterocytes. *J. Cell Biol.* **105**, 335–344.

Singer, R. H., Langevin, G. L., and Lawrence, J. B. (1989). Ultrastructural visualization of cytoskeletal mRNAs and their associated proteins using double-label *in situ* hybridization. *J. Cell Biol.* **108**, 2343–2353.

Stidwell, R. P., and Burgess, D. R. (1986). Regulation of intestinal brush border microvillus length during development by the G- to F-actin ratio. *Dev. Biol.* **114**, 381–388.

Stidwell, R. P., Wysolmerski, T., and Burgess, D. R. (1984). The brush border cytoskeleton is not static: *In vivo* turnover of proteins. *J. Cell Biol.* **98**, 641–645.

Takemura, R., Masaki, T., and Hirokawa, N. (1988). Developmental organization of the intestinal brush-border cytoskeleton. *Cell Motil. Cytoskel.* **9**, 299–311.

Tilney, L. G., and Cardell, R. R. (1970). Factors controlling the reassembly of the microvillus border of the small intestine of the salamander. *J. Cell Biol.* **47**, 408–422.

Tilney, L. G., and Mooseker, M. S. (1971). Actin in the brush-border of epithelial cells of the chicken intestine. *Proc. Natl. Acad. Sci. U.S.A.* **68**, 2611–2615.

Toner, P. G. (1968). Cytology of intestinal epithelial cells. *Int. Rev. Cytol.* **24,** 233–343.

Trier, J. S. (1963). Studies of small intestinal crypt epithelium. I. The fine structure of the crypt epithelium of the proximal small intestine of fasting humans. *J. Cell Biol.* **18,** 599–620.

van Dongen, J. M., Visser, W. J., Daems, W. Th., and Galjaard, H. (1976). The relation between cell proliferation, differentiation and ultrastructural development in rat intestinal epithelium. *Cell Tissue Res.* **174,** 183–199.

Weinmann, M. D., Allan, C. H., Trier, J. S., and Hagen, S. J. (1989). Repair of microvilli in the rat small intestine after damage with lectins contained in the red kidney bean. *Gastroenterology* **97,** 1192–1204.

West, A. B., Isaac, C. A., Carboni, J. M., Morrow, J. S., Mooseker, M. S., and Barwick, K. W. (1988). Localization of villin, a cytoskeletal protein specific to microvilli, in human ileum and colon and in colonic neoplasms. *Gastroenterology* **94,** 343–352.

Younes, M., Harris, A. S., and Morrow, J. S. (1989). Fodrin as a differentiation marker. Redistributions in colonic neoplasia. *Am. J. Pathol.* **135,** 1197–1212.

7

Development of the Chicken Intestinal Epithelium

Salim N. Mamajiwalla, Karl R. Fath, and David R. Burgess
Department of Biological Sciences
University of Pittsburgh
Pittsburgh, Pennsylvania 15260

I. Introduction
II. The Intestinal Epithelium
   A. Morphogenesis
   B. The Adult Crypt of Lieberkühn
III. The Cells of the Villus: Formation of the Brush Border
   A. Molecular Organization of the Mature Brush Border
   B. Enterocyte Development in Adult Crypts of Lieberkühn
   C. Maintenance of the Mature Brush Border
   D. Regulation of Microvillus Formation
IV. Regulation of Development
   References

I. Introduction

With the development of new techniques in cell and molecular biology the intestinal epithelium has emerged as one of the most fascinating model systems for cell biologists. Once the sole domain of gastroenterologists and physiologists, the intestinal epithelium has provided answers to some of the most fundamental questions in cell biology. The dynamic nature of the intestinal epithelium was first observed by Patzelt in 1882 (see Leblond and Messier, 1958, for references). The cells of the intestinal epithelium (Cheng and Leblond, 1974) are undergoing constant renewal: mitotically active cells, present at the base of the crypts (see Fig. 1), give rise to differentiated, nonproliferative cells that migrate along the length of the villus and are eventually lost into the intestinal lumen.

The brush border (BB), which constitutes the apical domain of the terminally differentiated absorptive cells, is elaborated by fine, finger-like projections termed microvilli. This modification allows the cell to efficiently perform its primary function of absorption and terminal digestion of nutrients. In addition to absorption, the plasma membrane of the microvilli contains many hydrolases and transporter molecules that aid in the terminal digestion of nutrients (see Holmes and Lobley, 1989, for review and references). These enzymes have also turned

Current Topics in Developmental Biology, Vol. 26
Copyright © 1992 by Academic Press, Inc. All rights of reproduction in any form reserved.
123

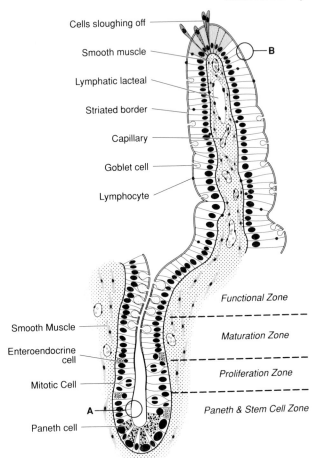

Cells sloughing off

Smooth muscle

Lymphatic lacteal

Striated border

Capillary

Goblet cell

Lymphocyte

B

Functional Zone

Smooth Muscle

Maturation Zone

Enteroendocrine
cell

Proliferation Zone

Mitotic Cell

A

Paneth & Stem Cell Zone

Paneth cell

Fig. 1 Crypt–villus continuum. A "typical" crypt–villus continuum of the small intestine, showing the relationship between one crypt of Lieberkühn and a villus. Stem cells residing in the proliferation zone give rise to either paneth, goblet, enteroendocrine, or absorptive cells. The latter three mature as they migrate up the villus and are eventually sloughed into the lumen. The paneth cells migrate to the base of the crypt. The detailed structures of the apical surfaces of the undifferentiated crypt cell (A) and the differentiated villus tip cell (B) are shown in Fig. 2.

out to be excellent markers for differentiation (Weiser, 1973a,b). The BB cytoskeleton is a rich source of a multitude of cytoskeleton proteins (for references, see below). The interaction of these proteins with one another has provided us with a generalized blueprint of the architecture of the cell. The information obtained thus far on the interactions of cytoskeleton proteins has given us an insight into how the cell might regulate its cytoskeleton under differing conditions. It is important to note, however, that much of this information has been obtained from *in vitro* studies and the role of many of these proteins *in vivo* has yet to be determined. While considerable progress has been made in understanding the constituents and their arrangement in the cytoskeleton of the absorptive cell very little is known about cytoskeleton assembly during development. In particular, our understanding of the cellular and molecular mechanisms regulating proliferation and differentiation of the stem cells that reside in the proliferative unit of the intestinal epithelium, the crypt of Lieberkühn, is very poor.

Our effort here is not to describe the characteristics of the components that make up the cytoskeleton of the BB; this information is presented only to shed light on the subject at hand. Rather, the focus of this article will be to describe what we currently know and do not know about the development of this highly dynamic system. We will also highlight some of the data that have emerged from our laboratory and other laboratories concerning the cellular and molecular events that might regulate the growth and differentiation of the cells in the crypts.

II. The Intestinal Epithelium

A. Morphogenesis

The morphogenesis of the intestinal epithelium in the embryonic chicken is simple in comparison to that of the rat and other mammals. The very early events leading to the formation of villi in the embryonic chicken were first observed and described by Hilton in 1902 and later by a number of investigators (Burgess, 1975; Coulombre and Coulombre, 1958; Grey, 1972). The duodenum of the 6-day embryo consists of a tube with a tiny lumen lined by pseudostratified epithelium. Prior to formation of definitive villi in the embryo the cells become cuboidal and the epithelium folds to form previllus ridges that run parallel to the length of the duodenum. Grey (1972) observed by scanning electron microscopy that villus formation is asynchronous and occurs in two phases: the first phase is completed by 13 days of incubation and consists of 16 previllus ridges; the second phase begins at day 16 and is completed by the fourth day after hatching. This second set of villi does not develop from previllus ridges but instead develops as tonguelike flaps between the first set of previllus ridges from the gut floor. The significance of this asynchronous development is not known but it should be pointed out that the villi within each set develops uniformly. Between 14 and 16 days of incubation, the first set of previllus ridges folds into a highly regular zig-zag pattern along its entire length. By day 17, definitive villi begin forming as a swelling of a distinct population of cells atop the crests of the ridges at the corner of each zig and zag. By day 18, this swelling of cells spreads along the length of each zig and zag and after 19 days the presumptive villi of the earlier stages take on the appearance of the finger-like villi characteristic of posthatching stages. The second set of villi that form from the floor of the duodenum at 16 days of embryogenesis grows very rapidly and is distinctly visible by the nineteenth day. At hatching, the only distinction between these late developing villi and the first set of villi is their height; the former are smaller. This distinction disappears, however, at the fourth day posthatching.

Previllus ridge formation from this point on is irregular, with 16 ridges being formed by day 13 of incubation. Ridge formation is possibly due to both active constrictions in groups of epithelial cells as well as induction by the underlying mesenchyme. The constrictions are most probably mediated by microfilaments in both the apical and basal domains of the epithelial cells (Burgess, 1975), because cytochalasin B, which disrupts microfilaments, prevents ridge formation. This microfilament-mediated constriction could be actomyosin based as myosin, tropomyosin, and actin, the elements required for contractility, are present in the adherens circumferential ring.

The sequential morphological events that lead to the formation of the adult intestinal epithelium of the fetal rat is much more complex and has been described in detail by Mathan *et al.* (1976) (see Figs. 14–16 in Mathan *et al.*, 1976, for excellent schematic diagrams). In the 15-day fetus (7 days before birth), the duodenum consists of a tube with a tiny lumen surrounded by stratified epithelial cells that are two to three layers thick and are in turn surrounded by mesenchymal cells. No microvilli are observed on the apical surface of cells exposed to the lumen. Long junctional complexes form between lateral membranes of both the apical cells and between cells in the deeper layers, and are different from the tight junctions of the adult epithelium. In the 17-day fetus (5 days before birth), the lumen is much larger and the epithelium consists of four to eight cell layers. A prominent basal lamina surrounds the most basal layer of epithelial cells. Lumenal cells express a few, small microvilli. These cells also contain large, lysosome-like structures, suggesting that these cells are beginning to degenerate. The long junctional complexes observed in the 15-day fetus are now abundant throughout the epithelium, especially in the deeper layers. More interestingly, small secondary lumena also form at these junctional complexes. As development proceeds, these secondary lumena enlarge as the cells surrounding these secondary lumena develop microvilli. By day 18 (4 days before birth), the main lumen is much larger due to the exfoliation of cells seen to contain lysosome-like vesicles in the 17-day fetus. Invagination of the underlying mesenchymal cells into the overlying epithelium results in the formation of epithelial folds. The invaginated mesenchyme probably ends up being the lamina propria in the adult. At this stage of development, the secondary lumena in the more superficial layers fuse to form cleftlike projections are lined with microvilli and the basal layer of epithelial cells are now more columnar in comparison to the cuboidal superficial epithelial cells. In the day-19 fetus, short villi are well formed and are lined with a single differentiated layer of epithelial cells composed of absorptive, goblet, and enteroendocrine cells. Prior to birth, true intestinal crypts of the adult are not present but develop after birth as shallow depressions from the flat intervillus epithelium (Dunn, 1967; Kammerad, 1942).

B. The Adult Crypt of Lieberkühn

Crypt development in the chicken has not been studied in as much detail as the development of the villi. Before the ninth day of incubation, mitotic cells are evenly distributed throughout the epithelium of the duodenum. After the ninth day, however, mitotic cells are gradually restricted to the bases of the villi until day 16, when the percentage of mitotic cells restricted to the bases of the villi increases sharply (Overton and Shoup, 1964). Almost no mitotic cells are seen on the distal portion of the villi after day 19 and by 2 days posthatching the mitotic cells are restricted to the crypts. The distribution of proliferating cells in the developing crypt of the rat is interesting in comparison to the chicken. Similar to the chicken, mitotic cells are initially present throughout the mucosa at all levels in the stratified epithelium. As secondary lumena begin to form, the mitotic cells are restricted to the lower layers of the epithelium and by 19 days of embryogenesis, when villi have formed, 50% of the mitotic cells are seen at the villus bases. By 22 days, mitotic cells are restricted to the crypt region. While villus morphogenesis in the chicken is different from that of the rat, it is interesting to note that the manner in which the proliferating cells are eventually restricted to the future crypt region is very similar. In both cases proliferating cells are initially present throughout the epithelium and are gradually restricted to the base of the villi at the same time the villi begin to form.

The unitarian theory of the origin of the epithelial cell types, first proposed in 1974 by Cheng and Leblond, has contributed immensely to our current understanding of the origin of the main cell types found in the intestinal epithelium. Tritium labeling studies from this classic work demonstrated that the four main types of cells, namely the villus columnar, oligomucous, enteroendocrine, and Paneth cells, that make up the intestinal epithelium arise by differentiation from the same progenitor cell. Their studies indicated that these four differentiated cell types arise from a stem cell pool that does not divide into functionally separate groups for each cell type but which is composed of multipotential stem cells. Thus, all the cells of the intestinal epithelium originate from the crypt-base columnar cells, which are believed to be the stem cells for the entire epithelium. Some of the daughter cells of these crypt-base columnar cells remain as stem cells, others migrate up the villus as they begin to differentiate.

In mice, the crypts are estimated to contain approximately 250 cells (Potten et al., 1982) while in rats and humans the number of cells per crypt is estimated to be between 650 and 900 (Wright et al., 1973, 1975). Anywhere between 50 and 80% of these cells are undergoing cell division at a steady state. The stem cells are thought to have a mean cell cycle time of 24 hr (Potten et al., 1982). Leblond and Cheng (1976) have calculated that the progeny arising from each stem cell can undergo three further divisions. These transitory proliferative cells are thought to have cell cycle times of about 13 hr (Potten et al., 1982).

The exact number and position of the stem cells are not known. Data from mouse aggregation chimeras (Ponder et al., 1985; Schmidt et al., 1988) suggest that the crypt is formed from one progenitor cell that may give rise to several stem cells that are responsible for the constant renewal of the epithelium. Tritium-labeling experiments and mathematical modeling of the crypt estimate that there are between 4 and 16 stem cells per crypt (Bjerknes and Cheng, 1981a,b,c; Loeffler *et al.*, 1986; also reviewed in Potten and Loeffler, 1987). Bjerknes and Cheng (1981a) argue that the stem cells are located in two zones of the crypt; in the base of the crypt among the Paneth cells and also in a zone at the fifth cell position (position 1 being the very bottom of the crypt; see Fig. 1). Based on labeling indices from tritium-labeling experiments together with complex mathematical modeling, Potten and colleagues place the stem cells above the highest Paneth cell, i.e., in a single zone at approximately the fifth position (Loeffler *et al.*, 1986; Potten and Hendry, 1983; also reviewed in Potten and Loeffler, 1987). Hence, cells located just below the stem cells would differentiate into Paneth cells and cells located just above the stem cell zone would differentiate into either goblet, enteroendocrine, or absorptive cells of the villus tip. It is important to point out that there is no compelling evidence as yet to favor either hypothesis.

As has already been discussed above, the progeny of the stem cells undergo three further divisions in the proliferative zone. As the cells migrate toward the villus they lose their potential to divide, acquiring the characteristics of functional cells as they enter the maturation zone. How migration occurs has been an area of intensive study. The many hypotheses proposed and the arguments for and against the mechanisms responsible for migration have been discussed in detail by Kaur and Potten (1986). Although the exact mechanism of migration is still not known, the data of Kaur and Potten surprisingly demonstrate that the crypts and villi behave independently with regard to cell migration. Cell migration in the crypts is an active process that seems to require protein synthesis and protein glycosylation. On the other hand, cell movements on the villus seems to be passive and dependent on the continued contraction of the smooth muscle cells in the underlying lamina propria. In any case, the role and regulation of the cytoskeleton in this migration have not been investigated.

III. The Cells of the Villus: Formation of the Brush Border

A. Molecular Organization of the Mature Brush Border

The cytoskeleton of the BB is easily described if it is considered to be composed of two domains, the microvillus core and the terminal web. However, it should be remembered that these two domains are intimately interconnected. Only a superficial description of the key cytoskeletal proteins of the BB is presented here because the purpose of this article is to deal with the development of the system. For a detailed presentation of the actin cytoskeleton of the BB the reader is referred to several recent reviews (Burgess, 1987; Mooseker, 1985; Louvard,

1989; Matsudaira and Janmey, 1988). The current view of the molecular organization of the mature BB is shown in Fig. 2B.

The surface of adult chicken enterocytes is covered by about 1200 hexagonally packed microvilli, which are about 1.7 μm in height and about 100 nm in diameter. The structure of the chicken microvillus is maintained by a bundle of about 15 actin filaments, which run from the tip of the microvillus to the base of the terminal web. The actin filaments are polarized with their plus or barbed (fast growing) ends at the microvillus tip. The bundle is maintained through the action of two well-characterized actin-bundling proteins, villin and fimbrin. Villin, a member of the calcium-regulated actin-binding protein family that includes gelsolin, acts in the absence of calcium to bundle actin filaments. In the presence of micromolar levels of calcium, villin severs filaments and serves as a barbed end-capping and actin-nucleating protein. Fimbrin is a member of the family of calcium-insensitive actin-bundling proteins that includes α-actinin, spectrin, and dystrophin (de Arruda et al., 1990). Fimbrin is likely the key actin-bundling protein of the core because it forms highly ordered unipolar bundles in vitro and is found in such highly ordered bundles as those in stereocilia. However, questions regarding its regulation and function during microvillus growth remain unanswered. While villin will bundle actin in vitro, it is still not clear whether these bundles are unipolar. Interestingly, villin is present in the very early endoderm (much earlier than fimbrin) and therefore may play a key actin-bundling role in the primitive microvilli formed early on the epithelium, as these bundles are not as highly ordered as in the adult. Consistent with these data is the recent work of Friederich et al. (1989) showing that transfected villin induces primitive microvillus formation in COS cells.

The other major actin-binding protein of the microvillus core is myosin I, first identified as the 110K protein (Matsudaira and Burgess, 1979). Myosin I, coupled with calmodulin, forms a double spiral of cross-bridges linking the actin bundle to the plasma membrane (Matsudaira and Burgess, 1982) and has recently been shown to be a member of the myosin I family of small myosins (lacking a tail) that is unable to form bipolar filaments but possesses a lipid-binding domain (Mooseker and Coleman, 1989; Hayden et al., 1990). The role of myosin I in the formed microvillus bundle is unknown.

One minor protein of the core that deserves mention is ezrin, first described by Bretscher (1983). Ezrin may play a key role in microvillar assembly based on some interesting recent experiments, in which ezrin targets domains of the plasma membrane where microvilli will form upon tyrosine phosphorylation by the epidermal growth factor (EGF) receptor (Bretscher, 1989). Interestingly, ezrin may also be a microtubule–actin cross-linker (Brigbauer and Solomon, 1989), although definitive data supporting this hypothesis are lacking. At present, we know too little about the effects of phosphorylation or other regulations to attribute any definitive function to ezrin in microvillus assembly.

The terminal web of the microvilli contains the base of the microvillus core,

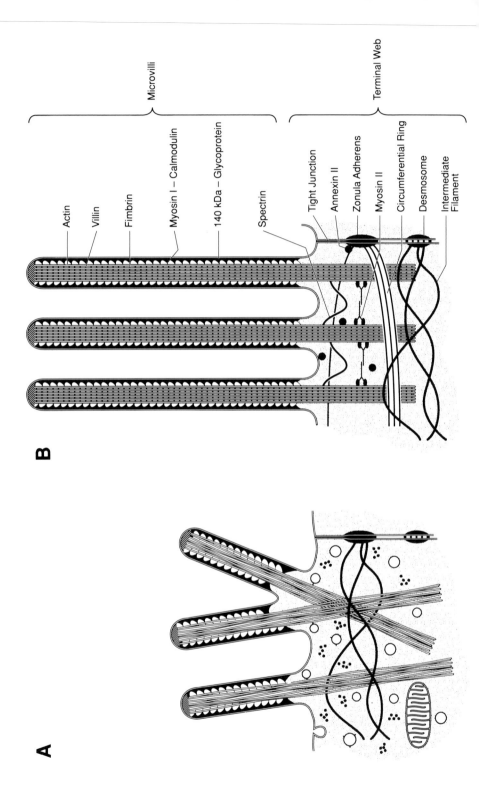

A

B

Microvilli

Terminal Web

Actin

Villin

Fimbrin

Myosin I – Calmodulin

140 kDa – Glycoprotein

Spectrin

Tight Junction

Annexin II

Zonula Adherens

Myosin II

Circumferential Ring

Desmosome

Intermediate Filament

now called the rootlet, enmeshed in a complex anastomosing network of filaments. In addition to the actin-associated proteins of the microvillus core, mentioned above, the rootlet also possesses tropomyosin and myosin II. The tropomyosin isoforms have been characterized and shown to have two primary effects on actin filaments. First, they protect actin from the calcium-activated severing action of villin (see reviews, and Burgess *et al.*, 1987). Second, they dramatically slow the rate of depolymerization of actin from the pointed end (Broschat *et al.*, 1989). The reasons for tropomyosin localization to the rootlet and absence from the microvillus core remain enigmatic, although there is speculation that caldesmon might play a role in this targeting (Bretscher and Lynch, 1985) because caldesmon enhances tropomyosin affinity for actin. In addition to tropomyosin the rootlet bundle also contains villin, fimbrin, and myosin I. The exact molar ratios of these with respect to actin are unknown but are thought to be less than in the microvillus core, partly because tropomyosin competes for the binding of villin and fimbrin to actin (Burgess *et al.*, 1987).

Immunolocalization and quick-freeze, deep-etch analyses localize BB spectrin, TW260/240, to the entire terminal web in a network of anastomosing filaments and myosin II in the lower portion of the terminal web (Hirokawa and Heuser, 1981; Hirokawa *et al.*, 1982, 1983a,b; Glenney and Glenney, 1983). It is likely that the myosin is in the form of bipolar filaments connecting rootlet bundles and that the spectrin forms similar contacts as well as attaching the actin to the plasma membrane.

The other major domain of actin cytoskeletal filaments is that forming the zonula adherens circumferential ring, as first noted by Hull and Staehelin (1979). Subsequent analysis of this ring demonstrated the presence of both actin and myosin II (Burgess and Prum, 1982) and also that the ring could be induced to constrict, much like a contractile ring, *in vitro* (Burgess, 1982). This confirms the hypothesis of Rodewald *et al.* (1976). Contraction of the ring is regulated by myosin II light chain phosphorylation (Keller and Mooseker, 1982; Broschat *et al.*, 1983) and likely plays a physiological role in affecting transepithelial resistance (Madara, 1989).

Fig. 2 The molecular organization of the chicken intestinal brush border. Some of the components and their likely arrangement in the apical cytoplasm of presumptive enterocytes in the intestinal crypt (A) and in mature villus enterocyte (B). (A) The microvilli of cells in the lower crypt express short microvilli that extend long rootlets into a rudimentary terminal web. Also note vesicles binding to the rootlets via proposed myosin I side arms as discussed in the text. (B) In comparison to cells of the crypt, mature enterocytes on the villus contain longer microvilli and shorter rootlets. The core bundle of actin filaments are now regularly arrayed and the terminal web is highly crosslinked. [(B) is adapted from Holmes and Lobley, 1989.]

B. Enterocyte Development in Adult Crypts of Lieberkühn

1. Ultrastructural Changes during Brush Border Assembly in the Crypt

Although the BB is assembled in the maturing cells of the adult crypt in much the same order as in the embryo, its formation in the adult crypts is much quicker (Brown, 1962; Trier and Rubin, 1965; Fath et al., 1990; Heintzelman and Mooseker, 1990). In fact, by the time the cells have migrated out of the proliferative zone into the maturation zone, the BB is nearly assembled. Initially the cells have domed apical surfaces that extend into the gut lumen. See Fig. 2A for a drawing of an immature crypt cell. These cells extend a few microvilli that arise from the center of the lumen, near the top of the dome. These microvilli are only 0.5 μm long and run at many angles relative to the cell surface. The microvillar rootlets are longer than the microvilli, and extend as much as 1 μm into the apical cytoplasm (Fath et al., 1990). As the enterocyte further develops, the number and length of the microvilli gradually increase and the rootlets shorten. The microvillus core actin filaments, although bundled in the early microvilli, become more regularly arrayed as the microvilli orient perpendicular to the cell surface. This increased order may reflect a change in the number or identity of the actin-binding cross-linking elements. When the enterocytes have migrated onto the villus the microvilli have nearly reached their mature length.

Because the total length of the core actin bundle (including the microvillus and rootlet) is approximately the same in the crypt and villus, we have proposed that the microvillus elongates by membrane addition at the microvillus base (Fath et al., 1990). Such addition would in effect lengthen the microvillus and shorten the rootlet. This proposal is consistent with our observations in the electron microscope (Fath et al., 1990). This membrane, we propose, may be supplied by the vesicles observed in the apical cytoplasm with myosin I on their cytoplasmic faces (Drenckhann and Dermietzel, 1988; unpublished observations). We have included several of these vesicles with attached myosin I in Fig. 2A. The vesicles with surface myosin I could bind to the rootlet and move toward the plus end of the actin filaments, which is the microvillus tip. When the vesicles reach the microvillus base, they may fuse with the plasma membrane and therefore link the myosin I to the actin core with the microvillus membrane.

The terminal web in the adult crypts develops in two phases. Initially the most immature cells contain a rudimentary terminal web with a well-formed zonula adherens circumferential ring. There are many polysomes and membranous vesicles scattered throughout the apical cytoplasm and approaching the lumenal membrane. In the final phase, near the end of maturation as the cells migrate onto the villus, the terminal web becomes highly cross-linked, excluding polyribosomes and membranous vesicles from the apical cytoplasm.

2. Expression of Crypt Cytoskeletal Proteins during Brush Border Assembly

In order to ascertain possible mechanisms for BB assembly, the distribution and levels of expression of the major actin-cytoskeletal proteins have been explored along the crypt–villus continuum. The reason for this is that if the expression of a certain protein(s) parallels morphological assembly then it may be a good candidate for a potential regulator of this assembly. Immunocytochemical studies show that immature cells in the proliferation zone of the chicken crypt that have not assembled a BB contain actin, villin, myosin II, tropomyosin, spectrin, and fimbrin at their apical membranes (Boller et al., 1988; Fath et al., 1990; Heintzelman and Mooseker, 1990). These studies suggest that the terminal web and microvilli are assembled in adult crypt epithelia from a preformed pool of cytoskeletal proteins concentrated in the apical cytoplasm. Myosin I, unlike the other actin-cytoskeleton proteins, is not restricted to the apical plasma membrane of immature crypt cells but is present diffusely throughout the apical cytoplasm. Electron microscopy shows, however, that the early microvilli contain the characteristic myosin I side arms that connect the core bundle of actin filaments with the microvillus plasma membrane (Fath et al., 1990). By the time the enterocytes migrate to the maturation zone of the crypt, myosin I is now concentrated at the plasma membrane (Fath et al., 1990; Heintzelman and Mooseker, 1990).

The suggestion from immunocytochemical studies that the major cytoskeletal proteins are expressed in their full abundance in cells prior to, or coincident with, BB assembly is supported by biochemical studies. Using an enzyme-linked immunosorbent assay (ELISA) we have determined the levels of the major actin-cytoskeletal proteins expressed in pure populations of enterocytes isolated from the crypt, lower villus, and upper villus (Fath et al., 1990). We observed no significant change in the levels of expression of actin, α-actinin, myosins I and II, spectrin, and tropomyosin along the crypt–villus continuum. We did detect a roughly twofold increase in villin levels between the crypt and lower villus cells that was also seen in adult mouse intestine (Boller et al., 1988).

3. Levels of Expression of Cytoskeletal Protein mRNAs during Enterocyte Maturation in the Adult Crypt

Studies have also explored correlations between changes in the expression of cytoskeletal protein mRNAs that parallel morphological development of the BB. Not surprisingly, the increase in villin protein levels during enterocyte maturation corresponds to a similar threefold increase in mRNA steady state levels in isolated chicken enterocytes when quantified on Northern blots (Fath et al., 1990) and in the mouse intestine using in situ hybridization (Ezzell et al., 1989). In isolated chicken crypt and villus enterocytes, we also detected a severalfold increase in the levels of calmodulin and tropomyosin mRNAs; however, the message levels of spectrin, myosin II, and actin remained constant (Fath et al., 1990). A similar constant level of actin mRNA along the crypt–villus continuum was also found in mice (Cheng and Bjerknes, 1989).

C. Maintenance of the Mature Brush Border

Although the lifespan of an enterocyte is only several days, the proteins composing the BB are constantly being renewed. In vivo pulse-labeling studies in chick villus enterocytes have shown a turnover time of several hours for myosin I, TW260/240, actin, villin, and fimbrin (Stidwell et al., 1984). Using shorter pulse times in pig intestinal explants, Cowell and Danielsen (1984) have measured a turnover rate for villin and myosin I with a half-life of at least 30 min. In addition to the cytoskeletal proteins, the intestinal BB membrane is quite dynamic, with the half-lives of both enzymes (Alpers, 1972; Hauri et al., 1985) and glycoproteins (Quaroni et al., 1979) being less than 1 hr. This dynamic state of the BB is further evidenced by the transient shortening of intestinal microvilli in cells treated with protein synthesis inhibitors (LeCount and Grey, 1972) or microvillar shortening and terminal web disruption in fasting animals (Misch et al., 1980).

Although the constituents of the BB are in a dynamic flux, there is tight regulation of the BB morphology. Quite strikingly, the length and number of microvilli and the number of core actin filaments in each microvillus are constant. Moreover, enterocytes at the same stage of maturation, whether in the adult crypts or in the embryo, exhibit microvilli of the same length. The final lengths of the mature microvillus in a single BB and in the duodenum of each animal are approximately the same. It has been proposed that microvillar length may be regulated by the G-actin:F-actin ratio (Stidwell and Burgess, 1986). These workers found that microvilli rapidly elongate in chicks 2–4 days after hatching. By determining the total length of the actin filaments in microvilli and the cell G-actin:F-actin ratio during development, they determined that an increase in G-actin probably drives assembly of more F-actin and thus causes microvillar elongation.

Although the proteins of the microvillus core are constantly turning over and microvillar length is constant, there must be fine regulation to maintain this structure. In vitro, actin monomers can be added to either end of the microvillus core, although addition occurs preferentially at the rootlet end (Mooseker et al., 1982). We have suggested that villin caps the filament ends at the tips of microvilli (Broschat et al., 1989) because the concentration of villin vastly exceeds the number of barbed filament ends, thus allowing polymerization at the rootlet end that is stabilized by tropomyosin. This idea is consistent with the experiments of Mooseker et al. (1982), who showed that actin monomers could add easily to the rootlet end *in vitro* but could only add to the barbed end under high EGTA, conditions that would remove villin from the barbed ends of actin filaments.

D. Regulation of Microvillus Formation

Although the signal for initiation of BB assembly has not been elucidated, a growing body of evidence indicates the importance of villin in early events in formation or stabilization of microvilli. The actin-binding protein villin has found favor because its interaction with actin can be regulated and it is expressed only in cells containing a BB. More direct evidence comes from the transient expression of villin in CV-1 monkey kidney fibroblasts (Friederich *et al.*, 1989) and the microinjection of purified villin into mouse NIH 3T3 fibroblasts (Franck *et al.*, 1990). In both studies, these cells, which do not normally express villin, were induced to form microvilli containing the exogenous villin. The normal actin cytoskeleton was disrupted as villin and other actin-binding proteins were incorporated into the microvilli. These studies also found that microvillus formation required both actin-binding domains of villin. When one of the actin-binding domains was removed, leaving only the Ca^{2+}-dependent actin severing, capping, and nucleation sites, the actin cytoskeleton was disrupted; however, no microvilli formed (Franck *et al.*, 1990). Further evidence for a important role for villin in microvillus formation comes from studies with a human colonic adenocarcinoma HT-29 cell line. These cells can be induced to differentiate into enterocytes and goblets cells by changing the media conditions (Dudouet *et al.*, 1987). As the differentiated cells are selected, the level of villin mRNA and protein increases 10-fold (Dudouet *et al.*, 1987). These studies taken together suggest the importance of villin in microvillus structure; however, there has been no direct evidence that villin alone can nucleate or induce the formation of the microvillus.

IV. Regulation of Development

Using isolated populations of undifferentiated crypt and differentiated villus tip cells from the adult intestinal epithelium, Burgess *et al.* (1989) demonstrated that proteins are not only phosphorylated on tyrosine in the adult chicken intestine but that the level of protein tyrosine phosphorylation is 15-fold higher in cells of the proliferative unit (the crypt) of the intestinal epithelium than in the differentiated, nondividing cells of the villus tip. Furthermore, most of the phosphotyrosine was detected on two proteins of 17 and 36 kDa. *In vitro* phosphorylation experiments with cytoskeletons demonstrated that both the 17- and 36-kDa proteins are phosphorylated on tryosine residues and are exclusively found on the cytoskeleton.

The fact that both these proteins can be phosphorylated *in vitro* indicates that the tyrosine kinase(s) must also be localized on the cytoskeleton and must be in intimate contact with these proteins, or alternatively must be in contact with an intermediate tyrosine kinase.

In apparent contrast with our results using isolated cells, previous studies using immunofluorescence (Takata and Singer, 1988) or Western blots of whole-tissue homogenates (Maher and Pasquale, 1988) did not detect phosphotyrosine-containing proteins in adult chicken intestine. Because crypts compose only a small percentage of the intestine, we think that inability to detect phospho-tyrosine containing proteins in the adult intestinal epithelium is most likely due to a lack of signal intensity rather than to the absence of tyrosine-phosphorylated residues.

What is the nature of the 36-kDa and the 17-kDa proteins that we observe being phosphorylated to a greater extent in the crypts than in the tips? Our preliminary data by two-dimensional sodium dodecyl sulfate-polyacrylamide gel electrophoresis (SDS-PAGE) and Western blots indicate that the 36-kDa protein might in fact be the predominant substrate for the protein-tyrosine kinase activity of pp60$^{v\text{-}src}$ as shown by Radke and Martin (1979). p36, also termed calpactin I or lipocortin II, is one member of a family of unique-calcium binding proteins that is not characterized by the EF-hand motif found in many of the traditional calcium-binding proteins such as calmodulin, troponin C, S-100, and the calbindins. The members of this family have been identified in a number of laboratories utilizing different approaches and have thus acquired a plethora of names. Recently this family of proteins has been retermed the annexins (Crumpton and Dedman, 1990). At present six annexins (reviewed in Burgoyne and Geisow, 1989; Crompton *et al.*, 1988; Hunter, 1989) have been isolated and cloned. Of these six annexins, p36, now classified as annexin II, has been the most studied because it is the predominant substrate of the oncogene pp60$^{v\text{-}src}$ and it is an actin-binding protein.

The use of monoclonal antibodies has shown that annexin II exists either as a soluble form or as a complex that is bound to the cytoskeleton (Zokas and Glenney, 1987). When bound to the cytoskeleton, annexin II is complexed with an 11-kDa light chain. Gel-filtration experiments show that this complex has a molecular weight of 85,000 and occurs as a heterotetramer in the form $(p36)_2(p11)_2$ (Gerke and Weber, 1984). It is not known whether the annexin II complex is bound to the cytoskeleton via p11 or some other polypeptide.

Recently, Ikebuchi and Waisman (1990) have been able to demonstrate *in vitro* bundling of F-actin by the annexin II complex at physiological levels of 0.1–2 μM free Ca^{2+}. Half-maximal binding of the annexin II complex occurred at approximately 0.18 μM free protein and is within the range reported for many of the other actin-binding proteins. Their data show that each annexin II complex is able to bind 1.5–2 actin molecules. Annexin II alone bundled F-actin to a lesser extent and at higher concentrations than the complex, suggesting that p11 might also play an important part in regulating the F-actin-bundling activity of annexin II. In the presence of the phospholipids phosphatidylserine and phosphatidylinositol the calcium requirement for actin binding is reduced to 10^{-6}–$10^{-5} M$ (J. Glenney, 1986; J. R. Glenney, 1985; Glenney et al., 1987). It has been demonstrated (Powell and Glenney, 1987) that binding of the annexin II complex to phospholipids can be achieved at free Ca^{2+} concentrations of $<10^{-8} M$, suggesting that the complex may be bound to phospholipids in vivo at physiological concentrations of Ca^{2+}. On Ca^{2+} binding, a conformational change in annexin II occurs, exposing one or more tyrosine residues (Gerke and Weber, 1985).

The ability of annexin II to bind phospholipids in the presence of calcium at levels found in resting cells implies that it may be involved in exocytosis via membrane–membrane interactions. Quick-freeze, deep-etch electron microscopy as well as immunocytochemistry strongly suggest that the annexin II complex is involved in the exocytosis of vesicles by cross-linking vesicles to the plasma membrane (Nakata et al., 1990). When the annexin II complex is localized by colloidal gold particles on cryo-ultrathin sections the complex is seen to be more closely associated with the plasma membrane than the actin cortical cytoskeleton.

Numerous studies (Gerke and Weber, 1984; Greenberg and Edelman, 1983; Nigg et al., 1983; Radke et al., 1983), including the ones mentioned above, have shown that annexin II is always localized on the cytoskeleton at the inner face of the plasma membrane. Interestingly, cells that do not express annexin II in vivo will express and accumulate annexin II in culture (Carter et al., 1986; Isacke et al., 1989; Lozano et al., 1989). Schlaepfer and Haigler (1990) have recently demonstrated that in human foreskin fibroblasts (HFF) annexin II levels remain unchanged in cultures induced to proliferate and concluded that annexin II does not play an important role in proliferation. It should be noted that none of the studies mentioned above has questioned the importance of p11 nor have these studies determined the level of tyrosine phosphorylation of annexin II. Although there may not be any change in the expression of annexin II, as suggested by Schlaepfer and Haigler (1990), its level of phosphorylation might be changing and this might be of physiological importance.

Indeed, while our results indicate that annexin II is phosphorylated 15-fold more in the cells of the crypt than of the villus tip, our preliminary data seem to indicate that the level of annexin II and its light chain associated with the cytoskeleton remain unchanged when crypt cells differentiate into villus tip cells. Hence, while the level of annexin II protein may not be important, the level of phosphorylated annexin II may be of significance. The hypothesis that annexin II may serve to stabilize the cortical cytoskeleton to the overlying plasma membrane should not be disregarded. The phosphorylated form of annexin II may serve to keep the cytoskeleton in a soluble form and the dephosphorylated form may stabilize the cytoskeleton. The annexin II tetramer $(p36)_2(p11)_2$ is bound to the cytoskeleton (Gerke and Weber, 1984; Glenney *et al.*, 1987; Glenney and Tack, 1985) and p11 is never found in the absence of annexin II (Zokas and Glenney, 1987), suggesting an important regulatory role for p11. The functional relationship between annexin II and p11 should be investigated. There is much that is still not known about annexin II. The biggest stumbling block is that we still do not know the true function(s) of annexin II nor do we know what the tyrosine phosphorylation of annexin II by p60^{src} means.

An obvious question related to our observation that there is more tyrosine-phosphorylated protein in the undifferentiated crypt cells concerns the kinase responsible for phosphorylating annexin II. We have recently shown that the *in vitro* tyrosine kinase activity of pp60^{c-src} is higher in crypt stem cell cytoskeletons (on average fourfold higher as measured by enolase phosphorylation or sevenfold higher as measured by autophosphorylation) than cytoskeletons from differentiated cells from the villus tip (Cartwright *et al.*, 1991). Fifty-three percent of this activity is associated with the crypt cytoskeleton and 13% is associated with the villus tip cytoskeleton. This is important in light of the fact that pp60^{c-src} activity is observed to be much higher in human colon carcinomas than in normal colonic mucosa (Cartwright *et al.*, 1989). Because the activation of pp60^{c-src} kinase activity has been shown to be an early event in colonic carcinogenesis (Cartwright *et al.*, 1990) it will be of interest to determine how pp60^{c-src} kinase activity is regulated as the stem cells and the proliferating cells of the crypt differentiate into the functional cells of the villus tip. One possibility is that the kinase activity of pp60^{c-src} is regulated by phosphorylation of Tyr-527, which is thought to decrease its activity (Cooper *et al.*, 1986). This phosphorylation might be regulated by a distinct Tyr-527 kinase (Okada and Nakagawa, 1989). We are currently pursuing these possibilities. In any case, we are hopeful that the intestinal system described here will provide some interesting and possibly unexpected insights into the regulation of gut development and in the etiology of colon carcinomas.

Acknowledgments

We would like to thank Drs. Gary Walker and Noellette Conway for their helpful suggestions in the preparation of the manuscript. We would also like to acknowledge Mr. Bill Johnston's assistance in preparing the figures. The original research reported herein was supported by NIH Grant DK 31643.

References

Alpers, D. H. (1972). The relation of size to relative rates of degradation of intestinal brush border proteins. *J. Clin. Invest.* **51,** 2621–2630.

Bjerknes, M., and Cheng, H. (1981a). The stem-cell zone of the small intestinal epithelium. I. Evidence from Paneth cells in the adult mouse. *Am. J. Anat.* **160,** 51–63.

Bjerknes, M., and Cheng, H. (1981b). The stem-cell zone of the small intestinal epithelium. II. Evidence from Paneth cells in the newborn mouse. *Am. J. Anat.* **160**, 65–75.

Bjerknes, M., and Cheng, H. (1981c). The stem-cell zone of the small intestinal epithelium. III. Evidence from columnar, enteroendocrine, and mucus cell in the adult mouse. *Am. J. Anat.* **160**, 77–91.

Boller, K., Arpin, M., Pringault, E., Mangeat, P., and Reggio, H. (1988). Differential distribution of villin and villin m-RNA in mouse intestinal epithelial cells. *Differentiation (Berlin)* **39**, 51–57.

Bretscher, A. (1983). Purification of an 80K protein that is a component of the isolated microvillus cytoskeleton, and its localization in nonmuscle cells. *J. Cell Biol.* **97**, 425–432.

Bretscher, A. (1989). Rapid phosphorylation and reorganization of ezrin and spectrin accompany morphological changes induced in A-431 cells by epidermal growth factor. *J. Cell Biol.* **108**, 921–930.

Bretscher, A., and Lynch, A. (1985). Identification and localization of immunoreactive forms of caldesmon in smooth and nonmuscle cells: A comparison with the distributions of tropomyosin and alpha actinin. *J. Cell Biol.* **100**, 1656–1663.

Brigbauer, E., and Solomon, F. (1989). A marginal band-associated protein has properties of both microtubule- and microfilament-associated proteins. *J. Cell Biol.* **109**, 1609–1620.

Broschat, K. O., Stidwell, R. P., and Burgess, D. R. (1983). Phosphorylation controls brush border motility by regulating myosin structure and association with the cytoskeleton. *Cell (Cambridge, Mass.)* **35**, 561–571.

Broschat, K. O., Weber, A., and Burgess, D. R. (1989). Tropomyosin stabilizes the pointed end of actin filaments by slowing depolymerization. *Biochemistry* **28**, 8501–8506.

Brown, A. (1962). Microvilli of the human jejunal epithelial cell. *J. Cell Biol.* **12**, 623–627.

Burgess, D. R. (1975). Morphogenesis of intestinal villi. II. Mechanism of formation of previllous ridges. *J. Embryol. Exp. Morphol.* **34**, 723–740.

Burgess, D. R. (1982). Reactivation of intestinal epithelial cell brush border motility. ATP-Dependent contraction via a terminal web contractile ring. *J. Cell Biol.* **95**, 853–863.

Burgess, D. R. (1987). The brush border: A model for structure, biochemistry, motility, and assembly of the cytoskeleton. *In* "Advances in Cell Biology" (K. R. Miller, ed.), Vol. 1, pp. 31–58. JAI Press Inc., Greenwich, Connecticut.

Burgess, D. R., and Prum, B. E. (1982). A re-evaluation of brush border motility. Calcium induces core filament solution and microvillar vesiculation. *J. Cell Biol.* **94**, 97–107.

Burgess, D. R., Broschat, K. O., and Hayden, J. (1987). Tropomyosin distinguishes between the two actin binding sites of villin and affects actin binding properties of other brush border proteins. *J. Cell Biol.* **104**, 29–40.

Burgess, D. R., Jaing, W., Mamajiwalla, S., and Kinsey, W. (1989). Intestinal crypt stem cells possess high levels of cytoskeletal-associated phosphotyrosine-containing proteins and tyrosine kinase activity relative to differentiated enterocytes. *J. Cell Biol.* **109**, 2139–2144.

Burgoyne, R. D., and Geisow, M. J. (1989). The annexin family of calcium-binding proteins. *Cell Calcium* **10**, 1–10.

Carter, C. V., Howlett, A. R., Martin, G. S., and Bissel, M. J. (1986). The tyrosine phosphorylation substrate p36 is developmentally regulated in embryonic avian limb and is induced in cell culture. *J. Cell Biol.* **103**, 2017–2024.

Cartwright, C. A., Kamps, M. P., Meisler, A. I., Pipas, J. M., and Eckhart, W. (1989). pp60^{c-src} activation in human colon carcinoma. *J. Clin. Invest.* **83**, 2025–2033.

Cartwright, C. A., Meisler, A. I., and Eckhart, W. (1990). Activation of the pp60^{c-src} protein kinase is an early event in colonic carcinogenesis. *Proc. Natl. Acad. Sci. U.S.A.* **87**, 558–562.

Cartwright, C. A., Eckhart, W., Mamajiwalla, S., Skolnick, S. A., and Burgess, D. R. (1991). In preparation.

Cheng, H., and Bjerknes, M. (1989). Asymmetric distribution of actin mRNA and cytoskeletal pattern generation in polarized epithelial cells. *J. Mol. Biol.* **210,** 541–549.

Cheng, H., and Leblond, C. P. (1974). Origin, differentiation and renewal of the four main epithelial cell types in the mouse small intestine. V. Unitarian theory of the origin of the four epithelial cell types. *Am. J. Anat.* **141,** 537–562.

Cooper, J. A., Gould, K. L., Cartwright, C. A., and Hunter, T. (1986). Tyr^{527} is phosphorylated in pp60^{c-src}: Implications for regulation. *Science* **231,** 1431–1434.

Coulombre, A. J., and Coulombre, J. L. (1958). Intestinal development. I. Morphogenesis of the villi and musculature. *J. Embryol. Exp. Morphol.* **6,** 403–411.

Cowell, G. M., and Danielsen, E. M. (1984). Biosynthesis of intestinal microvillar proteins: Rapid expression of cytoskeletal components in microvilli of pig small intestinal mucosal explants. *FEBS Lett.* **172,** 309–314.

Crompton, M. R., Moss, S. E., and Crumpton, M. J. (1988). Diversity in the lipocortin/calpactin family. *Cell (Cambridge, Mass.)* **55,** 1–3.

Crumpton, M. J., and Dedman, J. R. (1990). *Nature (London)* **345,** 212.

de Arruda, M. V., Watson, S., Lin, C-S., Leavitt, J., and Matsudaira, P. (1990). Fimbrin is a homologue of the cytoplasmic phosphoprotein plastin and has domains homologous with calmodulin and actin gelation proteins. *J. Cell Biol.* **111,** 1069–1079.

Drenckhahn, D., and Dermeitzel, R. (1988). Organization of the actin filament cytoskeleton in the intestinal brush border: A quantitative and qualitative immunoelectron microscope study. *J. Cell Biol.* **107,** 1037–1048.

Dudouet, D., Robine, S., Huet, C., Sahuquillo-Merino, C., Blair, L., Coudrier, E., and Louvard D. (1987). Changes in villin synthesis and subcellular distribution during intestinal differentiation of HT29-18 clones. *J. Cell Biol.* **105,** 359–369.

Dunn, J. S. (1967). The fine structure of the absorptive epithelial cells of the developing small intestine of the rat. *J. Anat.* **101,** 57–68.

Ezzell, R. M., Chafel, M. M., and Matsudaira, P. T. (1989). Differential localization of villin and fimbrin during development of the mouse visceral endoderm and intestinal epithelium. *Development (Cambridge, UK)* **106,** 407–419.

Fath, K. R., Obenauf, S. D., and Burgess, D. R. (1990). Cytoskeletal protein and mRNA accumulation during brush border formation in adult chicken enterocytes. *Development (Cambridge, UK)* **109,** 449–459.

Franck, Z., Footer, M., and Bretscher, A. (1990). Microinjection of villin into cultured cells induces rapid and long-lasting changes in cell morphology but does not inhibit cytokinesis, cell motility, or membrane ruffling. *J. Cell Biol.* **111,** 2475–2485.

Friederich, E., Huet, C., Arpin, M., and Louvard, D. (1989). Villin induces microvilli growth and actin redistribution in transfected fibroblasts. *Cell (Cambridge, Mass.)* **59,** 461–475.

Gerke, V., and Weber, K. (1984). Identity of p36K phosphorylated upon Rouse sarcoma virus transformed with a protein purified from brush borders; calcium-dependent binding to nonerythroid spectrin and F-actin. *EMBO J.* **3,** 227–233.

Gerke, V., and Weber, K. (1985). Calcium-dependent conformational changes in the 36-kDa subunit of intestinal protein I related to the cellular 36-kDa target of Rous sarcoma virus tyrosine kinase. *J. Biol. Chem.* **260,** 1688–1695.

Glenney, J. (1986). Phospholipid-dependent Ca^{2+} binding by the 36-kDa tyrosine kinase substrate (calpactin) and its 33-kDa core. *J. Biol. Chem.* **261,** 7247–7252.

Glenney, J. R., and Glenney, P. (1983). Spectrin, fodrin, and TW260/240: A family of related proteins lining the plasma membrane. *Cell Motil.* **3,** 671–682.

Glenney, J. R., Jr. (1985). Phosphorylation of p36 *in vitro* with pp60^{src}. Regulation by Ca^{2+} and phospholipid. *FEBS Lett.* **192,** 79–82.

Glenney, J. R., Jr., and Tack, B. F. (1985). Amino-terminal sequence of p36 and associated p10: Identification of the site of tyrosine phosphorylation and homology with S-100. *Proc. Natl. Acad. Sci. U.S.A.* **82,** 7884–7888.

Glenney, J. R., Jr., Tack, B., and Powell, M. A. (1987). Calpactins: Two distinct Ca^{++}-regulated phospholipid- and actin-binding proteins isolated from lung and placenta. *J. Cell Biol.* **104,** 503–511.

Greenberg, M. E., and Edelman, G. M. (1983). The 34kd pp60*src* substrate is located at the inner face of the plasma membrane. *Cell (Cambridge, Mass.)* **33,** 767–779.

Grey, R. D. (1972). Morphogenesis of intestinal villi. I. Scanning electron microscopy of the duodenal epithelium of the developing chick embryo. *J. Morphol.* **137,** 193–214.

Hauri, H. P., Sterchi, E. E., Bienz, D., Fransen, J. A., and Marxer, A. (1985). Expression and intracellular transport of microvillus membrane hydrolases in human intestinal epithelial cells. *J. Cell Biol.* **101,** 838–851.

Hayden, S. M., Wolenski, J. S., and Mooseker, M. S. (1990). Binding the brush border myosin I to phospholipid vesicles. *J. Cell Biol.* **111,** 443–451.

Heintzelman, M. B., and Mooseker, M. S. (1990). Assembly of the brush border cytoskeleton: Changes in the distribution of microvillar core proteins during enterocyte differentiation in adult chicken intestine. *Cell Motil. Cytoskel.* **15,** 12–22.

Hilton, W. A. (1902). The morphology and development of intestinal folds and villi in vertebrates. *Am. J. Anat.* **1,** 459–504.

Hirokawa, N., and Heuser, J. E. (1981). Quick-freeze, deep-etch visualization of the cytoskeleton beneath surface differentiations of intestinal epithelial cells. *J. Cell Biol.* **91,** 339–409.

Hirokawa, N., Tilney, L. G., Fujiwara, K., and Heuser, J. E. (1982). The organization of actin, myosin, and intermediate filaments in the brush border of intestinal epithelial cells. *J. Cell Biol.* **94,** 425–443.

Hirokawa, N., Cheney, R. E., and Willard, M. (1983a). Location of a protein of the fodrin-spectrin-TW260/240 family in the mouse intestinal brush border. *Cell (Cambridge, Mass.)* **32,** 953–965.

Hirokawa, N., Keller, T., Chasan, R., and Mooseker, M. S. (1983b). Mechanism of brush border contractility studied by the quick-freeze–deep-etch method. *J. Cell Biol.* **96,** 1325–1336.

Holmes, R., and Lobley, R. W. (1989). Intestinal brush border revisited. *Gut* **30,** 1667–1678.

Hull, B. E., and Staehelin, L. A. (1979). The terminal web. A re-evaluation of its structure and function. *J. Cell Biol.* **81,** 67–82.

Hunter, T. (1989). The Ca^{2+}/phospholipid-binding proteins of the submembraneous skeleton. *In* "Biology of Growth Factors. Molecular Biology, Oncogenes, Signal Transduction, and Clinical Implications" (J. E. Kudlow, D. H. MacLennan, A. Bernstein, A. I. Gotlieb, eds.), pp. 169–193. Plenum, New York.

Ikebuchi, N. W., and Waisman, D. M. (1990). Calcium-dependent regulation of actin filament bundling by lipocortin-85. *J. Biol. Chem.* **265,** 3392–3400.

Isacke, C. M., Lindberg, R. A., and Hunter, T. (1989). Synthesis of p36 and p35 is increased when U-937 cells differentiate in culture but expression is not inducible by glucocorticoids. *Mol. Cell. Biol.* **9,** 232–240.

Kammerad, A. (1942). Development of gastrointestinal tract of rat; histiogenesis of epithelium of stomach, small intestine and pancreas. *J. Morphol.* **70,** 323–351.

Kaur, P., and Potten, C. S. (1986). Effects of puromycin, cycloheximide and noradrenalin on cell migration within the crypts and on the villi of the small intestine. *Cell Tissue Kinet.* **19,** 611–625.

Keller, T., and Mooseker, M. S. (1982). Ca^{++}-calmodulin-dependent phosphorylation of myosin, and its role in brush border contraction *in vitro*. *J. Cell Biol.* **95,** 943–959.

Leblond, C. P., and Cheng, H. (1976). Stem cells in the small intestine of the mouse. *In* "Stem Cells of Renewing Cell Populations" (A. B. Cairnie, P. K. Lala, and D. G. Osmond, eds.), pp. 7–32. Academic Press, New York.

Leblond, C. P., and Messier, B. (1958). Renewal of chief cells and goblet cells in the small intestine as shown by radioautography after injection of thymidine-H^3 into mice. *Anat. Rec.* **132,** 247–259.

LeCount, T. S., and Grey, R. D. (1972). Transient shortening of microvilli induced by cycloheximide in the duodenal epithelium of chicken. *J. Cell Biol.* **53,** 601–605.

Loeffler, M., Stein, R., Wichmann, H.-E., Potten, C. S., Kaur, P., and Chwalinski, S. (1986). Intestinal cell proliferation. I. A comprehensive model of steady-state proliferation in the crypt. *Cell Tissue Kinet.* **19,** 627–645.

Louvard, D. (1989). The function of the major cytoskeletal components of the brush border. *Curr. Opin. Cell Biol.* **1,** 51–57.

Lozano, J. J., Silberstein, G. B., Hwang, S.-I., Haindl, A. H., and Rocha, V. (1989). Developmental regulation of calcium-binding proteins (calelectrins and calpactin I) in mammary glands. *J. Cell Physiol.* **138,** 503–510.

Madara, J. L. (1989). Loosening tight junctions. Lessons from the intestine. *J. Clin. Invest.* **83,** 1089–1094.

Maher, P. A., and Pasquale, E. B. (1988). Tyrosine phosphorylated proteins in different tissues during chick embryo development. *J. Cell Biol.* **106,** 1747–1755.

Mathan, M., Moxey, P. C., and Trier, J. S. (1976). Morphogenesis of fetal rat duodenal villi. *Am. J. Anat.* **146,** 73–92.

Matsudaira, P. T., and Burgess, D. R. (1979). Identification and organization of the components in the isolated microvillus cytoskeleton. *J. Cell Biol.* **83,** 667–673.

Matsudaira, P. T., and Burgess, D. R. (1982). Organization of the cross-filaments in intestinal microvilli. *J. Cell Biol.* **92,** 657–664.

Matsudaira, P. T., and Janmey, P. (1988). Pieces in the actin-severing protein puzzle. *Cell (Cambridge, Mass.)* **54,** 139–140.

Misch, P., Giebel, P., and Faust, R. (1980). Intestinal microvilli responses to feeding and fasting. *Eur. J. Cell Biol.* **21,** 264–279.

Mooseker, M. S. (1985). Organization, chemistry, and assembly of the cytoskeletal apparatus of the intestinal brush border. *Annu. Rev. Cell Biol.* **1,** 209–241.

Mooseker, M. S., and Coleman, T. R. (1989). The 110-kD protein–calmodulin complex of the intestinal microvillus (brush border myosin I) is a mechanoenzyme. *J. Cell Biol.* **108,** 2395–2400.

Mooseker, M. S., Pollard, T. D., and Wharton, K. A. (1982). Nucleated polymerization of actin from the membrane-associated ends of microvillar filaments in the intestinal brush border. *J. Cell Biol.* **95,** 222–233.

Nakata, T., Sobue, K., and Hirokawa, N. (1990). Conformational change and localization of calpactin I complex involved in exocytosis as revealed by quick-freeze, deep-etch electron microscopy and immunocytochemistry. *J. Cell Biol.* **110,** 13–25.

Nigg, E. A., Cooper, J. A., and Hunter, A. (1983). Immunofluorescent localization of a 39,000-dalton substrate of tyrosine protein kinases to the cytoplasmic surface of the plasma membrane. *J. Cell Biol.* **96,** 1601–1609.

Okada, M., and Nakagawa, H. (1989). A protein tyrosine kinase involved in regulation of pp60[c-src] function. *J. Biol. Chem.* **264,** 20886–20893.

Overton, J., and Shoup, J. (1964). Fine structure of cell surface specializations in the maturing duodenal mucosa of the chick. *J. Cell Biol.* **21,** 75–85.

Ponder, B. A. J., Schmidt, G. H., Wilkinson, M. M., Wood, M. J., Monk, M., and Reid, A. (1985). Derivation of mouse intestinal crypts from single progenitor cells. *Nature (London)* **313,** 689–691.

Potten, C. S., and Hendry, J. H. (1983). Stem cells in murine small intestine. *In* "Stem Cells" (C. S. Potten, ed.), pp. 155–199. Churchill-Livingston, Edinburgh and London.

Potten, C. S., and Loeffler, M. (1987). A comprehensive model of the crypts of the small intestine of the mouse provides insight into the mechanisms of cell migration and the proliferation hierarchy. *J. Theor. Biol.* **127,** 381–391.

Potten, C. S., Henry, J. H., Moore, J. V., and Chwalinski, S. (1982). Cytotoxic effects in gastrointestinal epithelium (as exemplified by small intestine). *In* "Cytotoxic Insult to Tissue" (C. S. Potten and J. H. Hendry, eds.), pp. 105–152. Churchill-Livingstone, Edinburgh and London.

Powell, M. A., and Glenney, J. R. (1987). Regulation of calpactin I phospholipid binding by calpactin I light-chain binding and phosphorylation by p60^{v-src}. *Biochem. J.* **247**, 321–328.

Quaroni, A., Wands, J., Trielstad, R. L., and Isselbacher, K. J. (1979). Epithelial cell cultures from rat small intestine. Characterization by morphologic and immunologic criteria. *J. Cell Biol.* **80**, 248–265.

Radke, K., and Martin, C. S. (1979). Transformation by Rous Sarcoma virus: Effects of *src* gene expression on the synthesis and phosphorylation of cellular polypeptides. *Proc. Natl. Acad. Sci. U.S.A.* **76**, 5212–5216.

Radke, K., Gilmore, T., and Martin, G. S. (1980). Transformation by Rous sarcoma virus: A cellular substrate for transformation-specific protein phosphorylation contains phosphotyrosine. *Cell (Cambridge, Mass.)* **21**, 821–828.

Radke, K., Carter, V. C., Moss, P., Dehazya, P., Schliwa, M., and Martin, G. S. (1983). Membrane association of a 36,000-dalton substrate for tyrosine phosphorylation in chicken embryo fibroblasts transformed by avian sarcoma viruses. *J. Cell Biol.* **97**, 1601–1611.

Rodewald, R., Newman, S. B., and Karnovsky, M. J. (1976). Contraction of isolated brush borders from the intestinal epithelium. *J. Cell Biol.* **70**, 541–545.

Schlaepfer, D. D., and Haigler, H. T. (1990). Expression of annexins as a function of cellular growth state. *J. Cell Biol.* **111**, 229–238.

Schmidt, G. H., Winton, D. J., and Ponder, B. A. J. (1988). Development of the pattern of cell renewal in the crypt-villus unit of chimaeric mouse small intestine. *Development (Cambridge, UK)* **103**, 785–790.

Stidwell, R. P., and Burgess, D. R. (1986). Regulation of intestinal brush border microvillus length during development by the G- to F-actin ratio. *Dev. Biol.* **114**, 381–388.

Stidwell, R. P., Wysolmerski, T., and Burgess, D. R. (1984). The brush border cytoskeleton is not static: *In vivo* turnover of proteins. *J. Cell Biol.* **98**, 641–654.

Takata, K., and Singer, S. J. (1988). Phosphotyrosine-modified proteins are concentrated at the membranes of epithelial and endothelial cells during tissue development in chick embryos. *J. Cell Biol.* **106**, 1757–1764.

Trier, J. S., and Rubin, C. E. (1965). Electron microscopy of the small intestine: A review. *Gastroenterology* **49**, 547–603.

Weiser, M. M. (1973a). Intestinal epithelial cell surface membrane glycoprotein synthesis. I. An indicator of cellular differentiation. *J. Biol. Chem.* **248**, 2536–2541.

Weiser, M. M. (1973b). Intestinal epithelial cell surface membrane glycoprotein synthesis. II. Glycosyltransferases and endogenous acceptors of the undifferentiated cell surface membrane. *J. Biol. Chem.* **248**, 2542–2548.

Wright, N. A., Watson, A. J., Morley, A., Appelton, D. R., and Marks, J. (1973). Cell kinetics in flat (avillous) mucosa of the human small intestine. *Gut* **14**, 701–710.

Wright, N. A., Al-Dewachi, H. S., Appelton, D. R., and Watson, A. J. (1975). Cell population kinetics in the rat jejunal crypt. *Cell Tissue Kinet.* **8**, 361–368.

Zokas, L., and Glenney, J. R., Jr. (1987). The calpactin light chain is tightly linked to the cytoskeletal form of calpactin I: Studies using monoclonal antibodies to calpactin subunits. *J. Cell Biol.* **105**, 2111–2121.

8

Developmental Regulation of Sarcomeric Gene Expression

Charles P. Ordahl
Department of Anatomy
University of California, San Francisco
San Francisco, California 94143

I. Introduction

Myogenesis has become a focal point of research over the past decade because, in many ways, differentiating striated muscle cells offer a window onto many aspects of developmental biology, from the molecular level to morphogenesis. During differentiation striated muscle precursor cells undergo profound transformations ultimately leading to the conversion of their cytoplasm into a crystalline array of the contractile proteins. There are examples of regulation of contractile protein expression at every conceivable level, including transcriptional, post-transcriptional RNA processing, translational control, and finally posttranslational modification. Nevertheless, research over the past 10 years clearly indicates that in (almost) every case the mechanisms that act to restrict expression of sarcomeric protein to striated muscle cells act principally at the level of gene transcription. Sarcomere assembly requires the simultaneous presence of a full

Current Topics in Developmental Biology, Vol. 26
Copyright © 1992 by Academic Press, Inc. All rights of reproduction in any form reserved.
145

complement of contractile proteins, thus the transcription of the contractile protein genes must be activated at the onset of myogenesis in a coordinated fashion.

This consideration has led to the general hypothesis that coregulated contractile protein genes will be found to possess common sequence elements *in cis* (cis elements) that act as binding sites for a relatively small number of regulatory factors *in trans* (trans factors). Thus, as the nucleotide sequences of contractile protein genes were determined via molecular cloning and DNA sequencing, a great deal of attention was paid to the search for sequence motifs that might be conserved in the regulatory domains of different muscle genes. Although a number of potentially interesting motifs were identified, none was found to be universally present in all muscle genes. Moreover, two other problems with this approach began to be apparent. First, the location of the crucial regulatory regions of different genes is not always obvious; and second, sequence motifs with bona fide regulatory functions may be so short and degenerate as to be unobvious to even the most diligent observer.

Thus it is necessary to employ functional tests to first find the regulatory regions of different contractile protein genes and then subsequently to identify the sequence motifs within those regulatory regions that are directly responsible for that regulation. In typical experiments, a marker gene is placed under the transcriptional control of a regulatory sequence element from a muscle gene and introduced into cells in culture. A muscle-specific regulatory element is expected to elicit a significantly higher level of expression of the marker gene product in differentiated muscle cells as compared to either undifferentiated muscle cells or nonmuscle cells. The *sine qua non* for demonstrating that such an element is a muscle-specific *determinant* (as opposed to elements merely required for activity) is its ability to convert an otherwise nonmuscle promoter into a muscle specific promoter. Having passed these criteria, the element is then subjected to site-directed mutagenesis and protein-binding experiments to determine the precise nucleotide sequence motifs that are responsible for the activity of the element.

The following discussion of the determinants of skeletal muscle-specific transcription is divided into two major categories: enhancer elements and promoter elements. According to this classification promoter elements are located very close (within about 200 nucleotides) to the transcriptional initiation site of the gene. Enhancer elements are typically located several hundreds or thousands of nucleotides either upstream or downstream from the site of transcription initiation. It must be emphasized, however, that while this division is useful for the purposes of review, the functional differences between enhancer and promoter elements may be more perceived than real. Indeed, in some cases both promoter and enhancer elements may be essentially equivalent except for their location.

Finally, because such experiments rely heavily on cell culture the vast majority of work on muscle gene expression has been analyzed in skeletal muscle cells, which are easily manipulated *in vitro* either as cell lines or as primary cultures.

For this reason the major portion of the work reported below on "muscle genes" reflects work in skeletal muscle. The final section, however, reports on the limited amount of work that has recently emerged to indicate how contractile protein genes are regulated in cardiac muscle cells.

II. Muscle-Specific Enhancers: A Unifying Theme Emerges

At present, the work with muscle-specific enhancers is giving us the most coherent picture of how muscle gene expression might be coordinated. This is primarily due to two factors: first, the recent realization that many, if not all, of the currently well-characterized muscle enhancers are probably governed by the same cis elements and trans factors. Thus, what we know about one muscle enhancer can, as a first approximation, be generalized to most if not all of the known muscle enhancers.

The second factor was the discovery that an entire family of muscle enhancer trans factors had already been cloned. Thus, virtually simultaneous with the discovery and characterization of a key cis element the cognate trans factor was available for study. The expression of these cloned factors has allowed detailed analysis of their interaction not only with target DNA sequences but also their protein–protein interactions with each other, and other trans factors with which they interact.

A. Common Structural Motifs of Muscle Enhancers

Several muscle genes are now known to possess muscle-specific enhancers (Donoghue *et al.*, 1988; Horlick and Benfield, 1989; Jaynes *et al.*, 1988; Konieczny and Emerson, 1987; Piette *et al.*, 1989; Sternberg *et al.*, 1989; Wang *et al.*, 1988; Yutzey *et al.*, 1989). The best characterized enhancer is that of the muscle creatine kinase gene, which is located over 1000 nucleotides upstream of the transcription initiation site (Horlick and Benfield, 1989; Jaynes *et al.*, 1988; Sternberg *et al.*, 1989). Two types of elements are known to be important for its function. The first, and best understood, are the MEF 1 (muscle enhancer factor 1) motifs first identified by Hauschka and co-workers (see Fig. 1) (Buskin and Hauschka, 1989). Gel shift and footprint experiments show that MEF 1 motifs bind nuclear factors present in muscle. Figure 1 shows the nucleotides which, when methylated, interfere with binding of MEF 1-binding factors. That the factors indeed recognize the MEF 1 motif is shown by the fact that mutations in the MEF 1 motif disrupt binding of the factors. Finally, mutations that disrupt MEF 1 binding also disrupt enhancer function. These three essential steps, therefore, complete the circle of arguments that prove that MEF 1 motifs, and the factors that bind to them, are necessary cis and trans components for the function of the muscle creatine kinase gene enhancer.

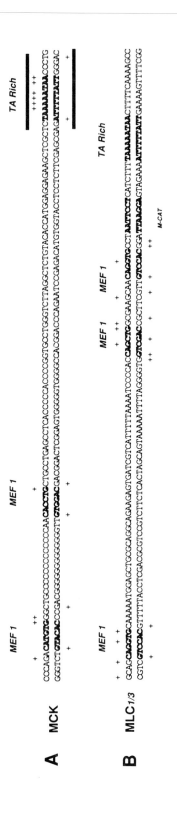

A MCK

MEF 1

MEF 1

CCAGACACTCCGGCTGCCCCCCCCCCCAACACTGCTGCCTGAGCCTCACCCCGGTGCCTGGGTCTTAGGCTCTGTACACCATGAGGAGAAGCTCGCTCTAAAAATAACCCCTG
GGGTCTGTACACCCGACGGGGGGGGGGGTTGACGACTCGGAGTGGGGTGGGGCCACGGACCCAGAATCCGAGACATGGTACCTCCTCTTCGAGCCGAGACAATTTTATTGGGAC

++   ++   +++ ++

+   +   +

TA Rich

B MLC1/3

MEF 1   MEF 1   MEF 1

GGAGCAGGTGCAAAAATGAGCTGGCAGGAGAAGAGTGATCGTCATTTTTAAAATCCCCACCACTGCGGCAAGCAACACGTGCCTAATTCCTCATCTTTTAAAAATAACCTTTTCAAAAGCC
CGTCGTCCACGTTTTTACCTCGACGGGTCCGTCTTCACTAGCAGTAAAAATTTTAGGGGTGTGACGCCCTTCGTTGTCCACGATTAAGATAGAAAATTTTATTGAAAAGTTTTCGG

+   + ++   +   + +   + + + + +   ++   ++

+   + +

M-CAT

TA Rich

C MEF 1 Motif

C NGG CA CC TG C N C
G NAA CA GG TG TT NG

**D Myfkin "Core"
Binding Site**

CANNTG

E AT-Rich Consensus

C TAAAAAATAAC CCC
T   TTT

Multiple MEF 1 motifs are also known to be present in the myosin light chain 1/3 gene enhancer (Donoghue *et al.*, 1988; Wentworth *et al.*, 1991; see Fig. 1). Footprint and mutational analysis of this enhancer show that the MEF 1 motifs are essential components of this enhancer (Wentworth *et al.*, 1991). Deletion and mutation experiments show that the redundancy of the MEF 1 motifs is an essential aspect of enhancer function because a minimum of two motifs is necessary for activity (Buskin and Hauschka, 1989; Weintraub *et al.*, 1990; Wentworth *et al.*, 1991).

The second component known to be important for enhancer function is an AT-rich segment that is located immediately downstream of the MEF 1 motifs of the both the muscle creatine kinase and myosin light chain 1/3 enhancers (Fig. 1). Deletion of this AT-rich segment substantially cripples the activity of the muscle creatine kinase enhancer (Gossett *et al.*, 1989; Horlick *et al.*, 1990), as well as that of the myosin light chain 1/3 enhancer (Wentworth *et al.*, 1991), indicating the AT-rich segment is an essential enhancer component. The AT-rich site of the muscle creatine kinase enhancer binds to a nuclear protein [dubbed MEF 2, after MEF 1, or TARP, for TA-rich recognition protein (Gossett *et al.*, 1989; Horlick *et al.*, 1990, respectively)]. Mutations of the AT-rich segment abolish protein binding and impair enhancer function (Horlick *et al.*, 1990). Detailed information regarding the proteins that bind to the TA-rich segments of these enhancers will be essential to determine how they collaborate with MyoD and related proteins to produce full enhancer effect.

B. MyoD and the Myfkin Family of Enhancer-Binding Proteins

While the Hauschka group was characterizing the MEF 1 motif and binding factor, independently at a nearby institution Weintraub and co-workers had discovered a "muscle determination" gene, dubbed MyoD, that can convert non-muscle cells to skeletal muscle cells (Davis *et al.*, 1987). Remarkably, MyoD was found to specifically bind the MEF 1 motif identified by the Hauschka group

---

Fig. 1 Structure of two muscle-specific enhancers and regulatory motifs. (A) The sequence of the enhancer of the mouse creatine kinase gene (Buskin and Hauschka, 1989; Jaynes *et al.*, 1988), which is similar in organization and sequence to that of the rat (Horlick and Benfield, 1989; Horlick *et al.*, 1990). Plus signs over and under the sequence indicate the nucleotides that, when methylated, interfere with factor binding (Buskin and Hauschka, 1989; Gossett *et al.*, 1989; Horlick *et al.*, 1990; Lassar *et al.*, 1989) while over- and underlining indicate nucleotides protected from chemical or DNase I cleavage (Gossett *et al.*, 1989). The core MEF 1- and AT-rich sequences are shown in boldface type. (B) The sequence of the enhancer of the myosin light chain 1/3 gene (Donoghue *et al.*, 1988; Wentworth *et al.*, 1991). A sequence motif related to the M-CAT motif is noted. All other notations as in (A). (C) Consensus for the MEF 1 motif as defined by Buskin and Hauschka (1989). (D) Core of MEF 1 sequence, which is the binding site for myfkins (Lassar *et al.*, 1989). (E) Consensus of AT-rich element as defined by Gossett *et al.* (1989).

Charles P. Ordahl

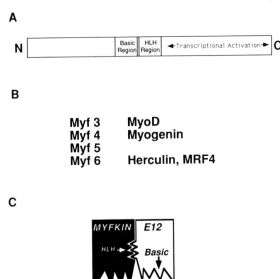

Fig. 2 The Myfkin family. (A) Basic structure of Myfkins based on the structure of MyoD as established by Weintraub and collaborators (Davis *et al.*, 1987, 1990; Lassar *et al.*, 1989, 1991; Tapscott *et al.*, 1988). The location of the transcriptional activation domain is based on Braun *et al.* (1990b). (B) Synonyms for mammalian myfkins (Myfkins isolated from nonmammals are related to one or more of these). The Myf series, originally proposed by Braun *et al.* (1989a, 1990a), affords a systematic array of the Myfkin family and facilitates comparisons. The source of the synonyms are MyoD (Davis *et al.*, 1987), myogenin (Wright *et al.*, 1989), Herculin (Miner and Wold, 1990), and MRF4 (Rhodes and Konieczny, 1989). References for nonmammalian Myfkins are de la Brousse and Emerson (1990), Edmondson and Olson (1989), Hopwood *et al.* (1989), Lin *et al.* (1989), Rhodes and Konieczny (1989), and Scales *et al.* (1990). (C) Schematic representation of the interaction of a Myfkin–E12 heterodimer and the core-binding sequence of the MEF 1-binding site (for details see text).

(Lassar *et al.*, 1989). Soon after the Weintraub group discovery, several other muscle "determination" genes were isolated (Braun *et al.*, 1989a, 1990a; de la Brousse and Emerson, 1990; Edmondson and Olson, 1989; Hopwood *et al.*, 1989; Lin *et al.*, 1989; Miner and Wold, 1990; Pinney *et al.*, 1988; Rhodes and Konieczny, 1989; Scales *et al.*, 1990; Wright *et al.*, 1989) that are closely related at the primary protein sequence level to MyoD (see Fig. 2). In terms of what is currently known about their molecular/biochemical properties, including binding to the MEF 1 motif, all of the different muscle "determination" genes appear to be essentially equivalent (for review, see Olson, 1990). Moreover, multiple Myfkins may be simultaneously present in the same cell and/or tissue (Sassoon *et al.*, 1989; M. Buckingham, 1990, personal communication) so that their differential roles are not yet entirely clear. The diverse nomenclature of these factors therefore masks what, at present, appears to be a high degree of functional similarity. For the sake of simplicity, therefore, members of this gene family will be referred to as "Myfkins," a term which incorporates the comprehensive Myf nomenclature (see Fig. 2) proposed by Arnold and co-workers (Braun *et al.*, 1989a) with the kinship relationship between the different members of this gene family among vertebrates and invertebrates.

Myfkins bind to the palindromic MEF 1 motif as dimers; homodimers bind weakly but heterodimers, formed with the ubiquitous transcription factor E12, bind very efficiently (Braun *et al.*, 1990b; Brennan and Olson, 1990; Lassar *et al.*, 1989, 1991) and it is probably a Myfkin–E12 dimer that functions *in vivo* (Lassar *et al.*, 1991). Homo- or heterodimerization occurs via an HLH (helix–loop–helix) domain that resides immediately adjacent to a stretch of basic amino acids that are responsible for DNA binding (see Fig. 2). Recent experiments indicate that a transcriptional activation domain resides toward the Myfkin carboxy-terminal end (Braun *et al.*, 1990b). Thus the current view is that for an enhancer to function, two or more MEF motifs must each be bound by a Myfkin–E12 heterodimer. How the putative transcriptional activation domain exerts its effect on the transcription initiation complex (possibly in concert with the factors binding to the TA-rich sequence) is at present unknown.

C. Myfkins as Myogenic Determination Genes

Although the discovery of Myfkins has provided the most concrete handle yet available for understanding muscle gene regulation, they pose interesting new problems that were previously unanticipated. Most intriguing is the ability of Myfkins to induce nonmuscle cells to undergo myogenic differentiation. A discussion of these properties falls outside the scope of this article and an excellent review of this subject has recently been published (Olson, 1990). Nevertheless, a key question is how the muscle determination properties of Myfkins are related to their properties as transcriptional regulators. It is well established that it is the DNA-binding site of MyoD that carries the "recognition code" for myogenic determination, suggesting that it is the site-specific DNA binding that is the essential aspect of both determination and transcriptional activation (Davis *et al.*, 1990). Surprisingly, however, some MyoD binding domain mutations that result in a loss of myogenic determination function do not disrupt binding to the core MEF 1 motif (Davis *et al.*, 1990). It would seem unlikely, therefore, that Myfkins induce myogenesis by simply turning on the muscle gene battery directly. Myfkins may induce myogenesis by way of interaction with a sequence motif that is distinct from MEF 1 and for which they have low affinity. This would also explain the necessity of high level of exogenous Myfkin expression to induce myogenesis. Determination of the various DNA sequences and proteins with which Myfkins interact should clarify the role this interesting family of proteins plays in both myogenesis, in general, and transcriptional activation of muscle genes, in particular.

D. Are There Other Muscle Enhancers?

The known enhancers of other genes, such as that encoding acetylcholinesterase (Piette *et al.*, 1990; X.-M. Wang *et al.*, 1990; Y. Wang *et al.*, 1988), cardiac troponin T (Mar *et al.*, 1988), and skeletal fast troponin I (Konieczny and Emerson, 1987; Yutzey *et al.*, 1989), all contain multiple MEF 1 sites that are likely to function in a manner similar to that established for the creatine kinase and myosin light chain genes described above. That the MEF 1 sites and Myfkins govern the promoter/enhancer of the gene encoding the α subunit of the acetyl-choline receptor gene has been recently established (Piette *et al.*, 1990, and see below). In addition, the promoter/enhancer of the gene encoding the δ subunit of the acetylcholine receptor is also likely to be governed in a similar manner, suggesting a possible mechanism of coordination between the genes encoding this complex receptor (Wang *et al.*, 1990). Thus, evidence is mounting that MEF 1–Myfkin-dependent enhancers are a major class of muscle specific determinants and, as such, represent the first achievement of a unifying control system for muscle gene activation. Nevertheless, the recent characterization of a Myfkin-independent muscle enhancer (Thompson *et al.*, 1991) suggests that greater diversity of enhancer mechanism may be found as more genes are analyzed. This would be more in accord with the multiplicity of cis elements and trans factors governing muscle promoters, as discussed below.

III. Muscle Promoters: A Diversity of Mechanisms

Unlike the unifying convergence in the study of muscle specific enhancers, the study of muscle-specific promoter determinants appears to become more di-vergent as more promoters are analyzed. A substantial number of muscle pro-moters is currently under study and many have now been shown to function in a muscle-specific fashion and conserved sequence motifs have been identified as possible, and even probable, muscle-specific determinants. Three different, con-served sequence motifs have been strongly implicated in the regulation of four different muscle promoters. For each, criteria comparable to those discussed above for the MEF 1 motif have been used to establish their function, i.e., (1) the DNA element has been shown to be capable of conferring muscle-specific ex-pression to a heterologous, nonmuscle promoter; (2) site-directed mutations within the conserved sequence motif have been shown to disrupt promoter func-tion; (3) the conserved sequence motif has been shown to bind a trans factor; and (4) the site-directed mutations that disrupt promoter function also disrupt binding of the trans factor. Taken together, these observations convincingly demonstrate that the sequence motif is an essential component of a muscle-specific determi-nant that exerts its effect on the transcriptional machinery via a sequence-specific interaction with a trans factor.

Interestingly, each of the four promoters contains a minimum of two of the three conserved motifs under discussion although, in all cases save one, only one motif appears to play a dominant role in each promoter. Thus, even when the discussion is limited to only four promoters, at least three different cis and trans determinant systems are evident. It should be kept in mind, therefore, that among the large number of interesting promoters not discussed here an even higher level of mechanistic diversity is likely to emerge.

A. Nonspecific Promoter Motifs

Figure 3 shows the nucleotide sequences of the promoter regions of the genes encoding skeletal actin, cardiac actin, cardiac troponin T, and the α subunit of the acetylcholine receptor. There is similarity in their overall organization, both in terms of sequence motifs conserved among muscle genes and in terms of canonical sequence elements common to many eukaryotic genes that are thought to be "nonspecific" promoter elements. For example, each has a nominal TATA box motif located within 20–30 nucleotides of the transcription initiation site (located at the far right of the figure). TATA box motifs are generally believed to play a role in the formation of the transcription initiation complex and, where analyzed, the nominal TATA boxes of muscle genes appear to serve roles similar to those of their nonmuscle counterparts (Bergsma *et al.*, 1986). It is interesting to note, in this regard, that nominal TATA box motifs of specific types of genes tend to be conserved over long evolutionary distances. It is possible, therefore, that such conserved TATA motifs also play specialized roles in the activity of each type of gene, although the limited experiments testing this notion have not reached such conclusions (Nikovits *et al.*, 1990). At least one muscle-specific promoter appears not to require a TATA motif for its function (Wang *et al.*, 1990).

B. GC-Rich Domains

Upstream from the nominal TATA box regions of muscle promoters reside sequence domains that are rich in G and C residues (see Fig. 3). Within such GC-rich domains are usually found one or more copies of the consensus sequence of the binding site of the transcription factor Sp-1 (Kadonaga *et al.*, 1986). Do these nominal Sp-1-binding sites actually function for Sp-1 binding or are they simply an accident of the GC-rich domain? The answer is not uniformly clear. Mutation of the nominal Sp-1 motif in the cardiac troponin T promoter has a modest deleterious effect on expression (Mar and Ordahl, 1990) while, conversely, that of the distalmost nominal Sp-1 motif in the skeletal actin promoter actually stimulates expression (Chow and Schwartz, 1990). On the other hand, Sp-1 binding has been strongly implicated as being necessary for activity of the

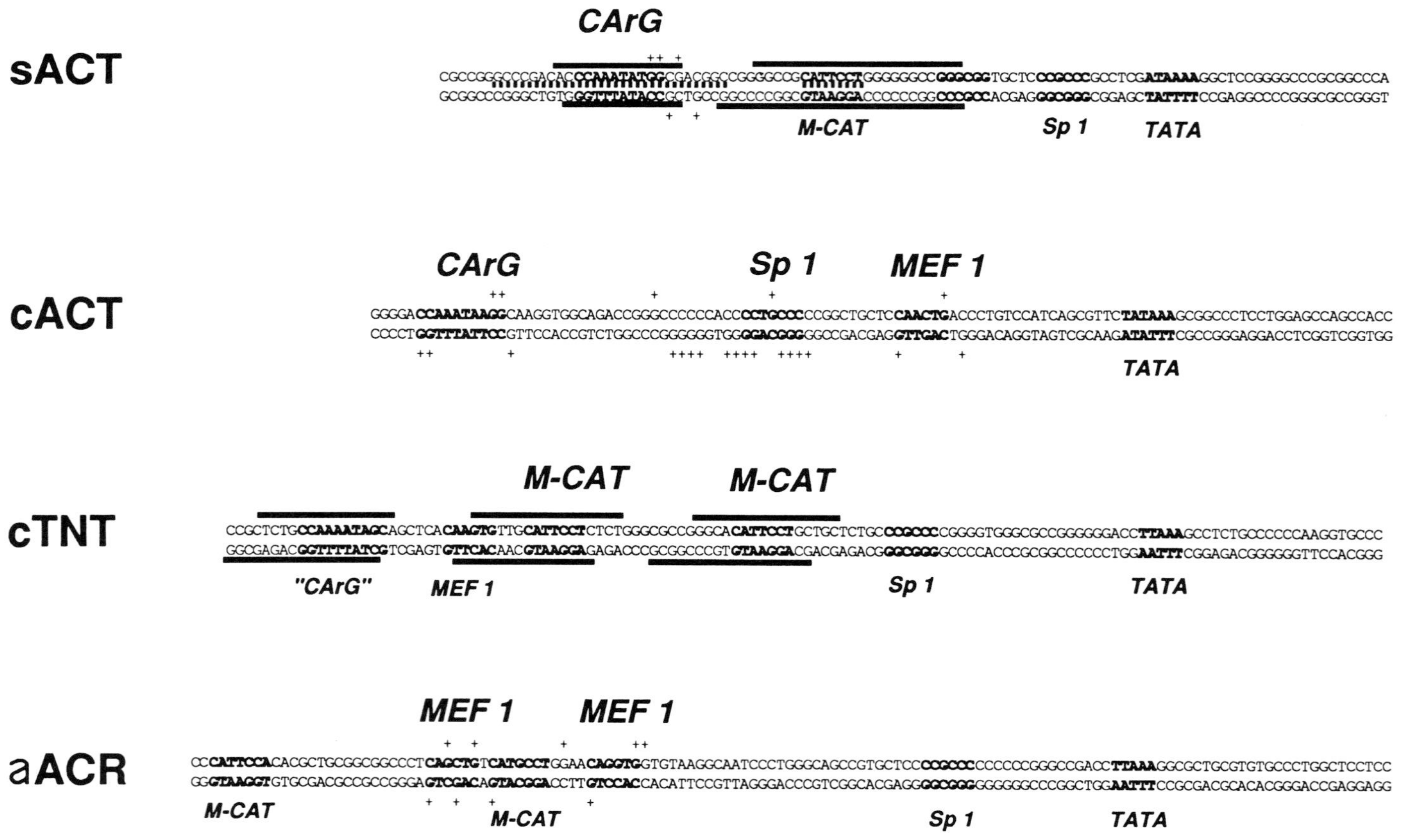

sACT
CArG
CGCCGGGCCCGACACCCAAATATGGCGACGGCCGGGGCCGCATTCCTGGGGGGCCGGGCGGTGCTCCCGCCCGCCTCGATAAAAGGCTCCGGGGCCCGCGGCCCA
GCGGCCCGGGCTGTGGGTTTATACCGCTGCCGGCCCCGGCGTAAGGACCCCCCGGCCCGCCACGAGGGCGGGCGGAGCTATTTTCCGAGGCCCCGGGCGCCGGGT
M-CAT
Sp 1
TATA
cACT
CArG
Sp 1
MEF 1
GGGGACCAAATAAGGCAAGGTGGCAGACCGGGCCCCCACCCCTGCCCCCGGCTGCTCCAACTGACCCTGTCCATCAGCGTTCTATAAAGCGGCCCTCCTGGAGCCAGCCACC
CCCCTGGTTTATTCCGTTCCACCGTCTGGCCCGGGGGTGGGGACGGGGGCCGACGAGGTTGACTGGGACAGGTAGTCGCAAGATATTTCGCCGGGAGGACCTCGGTCGGTGG
TATA
cTNT
M-CAT
M-CAT
CCGCTCTGCCAAAATAGCAGCTCACAAGTGTTGCATTCCTCTCTGGGCGCCGGGCACATTCCTGCTGCTCTGCCCGCCCCGGGGTGGGCGCCGGGGGACCTTAAAGCCTCTGCCCCCCAAGGTGCCC
GGCGAGACGGTTTTATCGTCGAGTGTTCACAACGTAAGGAGAGACCCGCGGCCCGTGTAAGGACGACGAGACGGGCGGGGCCCCACCCGCGGCCCCCCTGGAATTTCGGAGACGGGGGGTTCCACGGG
"CArG"
MEF 1
Sp 1
TATA
aACR
MEF 1
MEF 1
CCCATTCCACACGCTGCGGCGGCCCTCAGCTGTCATGCCTGGAACAGGTGGTGTAAGGCAATCCCTGGGCAGCCGTGCTCCCCGCCCCCCCCCGGGCCGACCTTAAAGGCGCTGCGTGTGCCCTGGCTCCTCC
GGGTAAGGTGTGCGACGCCGCCGGGAGTCGACAGTACGGACCTTGTCCACCACATTCCGTTAGGGACCCGTCGGCACGAGGGGCGGGGGGGGGCCCGGCTGGAATTTCCGCGACGCACACGGGACCGAGGAGG
M-CAT
M-CAT
Sp 1
TATA

cardiac actin promoter (Sartorelli *et al.*, 1990). Regardless of these apparent differences of effects, it is generally assumed that because Sp-1 is ubiquitous in its occurrence it is unlikely to play a role as a muscle-specific determinant. The conservation of GC-rich domains in muscle promoters suggests a functional significance and Sp-1 might collaborate with other factors to interact with these regions. It is interesting to note, in this regard, that the cardiac actin Sp-1 motif lies within a larger sequence element that is strongly conserved in the cardiac actin genes of mouse and humans (Sartorelli *et al.*, 1990).

C. Muscle-Specific Promoter Determinants

As indicated above, functional tests now implicate as least three different motifs as muscle-specific determinants; CArG motifs, M-CAT motifs, and MEF 1 motifs. Each of the four promoters shown in Fig. 3 contains at least two, and in one case all, of these motifs. Surprisingly, however, each promoter seems to rely principally on only one motif for its function in skeletal muscle cells. In each case, the motif(s) required for activity are indicated above the sequence of the promoter, while motifs that are not apparently required (or are believed to be simply canonical promoter motifs) are indicated below the sequence. A brief description of each motif, and the promoter(s) in which it was characterized, is presented below.

D. CArG Motifs and SRF

Among the first muscle gene promoters demonstrated to contain muscle-specific determinants were those of the sarcomeric actin genes. Early sequence analysis of actin genes identified sequence motifs in the skeletal actin gene promoters that have been conserved for hundreds of millions of years (Minty and Kedes, 1986; Nudel *et al.*, 1985; Ordahl and Cooper, 1983). One of these is a 20-nucleotide sequence motif that is conserved at 19 positions between birds and mammals (Fig. 1). The core of this 19-mer motif contains the sequence CCAAATATGG, which is closely related to a similar motif (CCAAATAAGG) in the cardiac actin

---

Fig. 3 Nucleotide sequence of the promoter regions of the chicken skeletal actin gene (s*ACT;* Chow and Schwartz, 1990), human cardiac actin gene (c*ACT;* Sartorelli *et al.*, 1990), chicken cardiac troponin T gene (Mar and Ordahl, 1990), and chicken α-acetylcholine receptor gene (Piette *et al.*, 1989). Plus signs indicate nucleotides that, when methylated, disrupt binding of the cognate trans-acting factor. Over- and underlining indicate regions protected from DNase I. Core nucleotide sequences for the various muscle specific motifs are shown in bold face. Sequences known to be important for the expression of the promoter are indicated above the nucleotide sequence while those that are dispensable are written below. Dots between upper and lower strands of sACT promoter indicate sequence elements conserved between birds and mammals (Ordahl and Cooper, 1983).

promoter. Related motifs with the general sequence $CC(A/T)_6GG$ (widely referred to by the phoneme "CArG") are also found in multiple copies upstream of both the cardiac and skeletal actin genes (Grichnik *et al.*, 1986; Minty and Kedes, 1986). Functional analyses showed that at least one CArG box is required for activity of both promoters (Grichnik *et al.*, 1986; Minty and Kedes, 1986; Mohun *et al.*, 1989) and that any one of the several CArG boxes is sufficient. Recent work with an embryonic myosin light chain promoter indicates that it too may be governed by a CArG-like motif (Uetsuki *et al.*, 1990). The CArG motifs can act as determinants of muscle specificity because multiple CArG motifs confer muscle-specific activity when linked to a heterologous, cell-nonspecific promoter (Tuil *et al.*, 1990).

The sequence of the CArG box motif is similar to that of the serum response element (SRE), which is responsible for up-regulation of the c-*fos* promoter in response to serum (Treisman, 1986). The SRE acts via binding of a transcription factor known as the serum response factor (SRF). Footprint and gel shift analyses show that the CArG motifs of the actin promoters bind in a sequence-specific fashion to nuclear proteins of muscle SRF (Boxer *et al.*, 1989b). Mutations of the CArG motif that disrupt actin promoter function also disrupt SRF binding, suggesting that SRF (and/or other factors closely related to SRF) activates actin gene transcription (Boxer *et al.*, 1989a; Miwa *et al.*, 1987; Mohun *et al.*, 1989; Walsh, 1989). It is puzzling, however, that a ubiquitous factor such as SRF could exert a muscle-specific effect on the actin promoter. Is it simply a question of context, that is, do the sequences surrounding the CArG motif play a role in its activity? While attractive, this explanation is not sufficient to explain how the c-*fos* CArG motif (SRE) can activate a heterologous promoter in a muscle-specific fashion (Tuil *et al.*, 1990). At least two other possibilities remain. First, muscle specificity may be conferred by a quantitative effect if muscle cells possess larger quantities of SRF than other cell types. Second, and more attractive, is the possibility that muscle cells possess other, cell-specific factors that also recognize the CArG motif and activate the actin promoters. Gel shift experiments suggest that other CArG-binding factors may exist (Walsh, 1989; Walsh and Schimmel, 1987), which could explain how CArG motifs exert their cell-specific effects.

E. MEF 1 Motifs and Myfkins

Recent work indicates that MEF 1 motifs, which were discussed above as enhancer motifs, also serve as important muscle-specific determinants in two different muscle promoters. The distal region of the promoter of the gene encoding the α subunit of acetylcholine receptor contains two MEF 1 sites with a center-to-center spacing of 18 nucleotides (see Fig. 3). Site-directed mutation of either site impairs the activity of this promoter in skeletal muscle cells while mutation of both sites essentially abolishes activity (Piette *et al.*, 1990). This dependence on two MEF 1 sites is in accord with a similar dependence of muscle enhancers on two MEF 1 sites (see above) and, indeed, the segment of the acetycholine receptor promoter containing the MEF 1 sites can act as a muscle-specific enhancer when juxtaposed to a nonmuscle promoter (Piette, *et al.*, 1989; Wang *et al.*, 1988). Recent reports indicate MEF 1 motifs may also govern the promoter of the gene encoding the δ subunit of the acetylcholine receptors (Wang *et al.*, 1990).

As discussed above, the MEF 1 motif is known to be a binding site for the Myfkin family of muscle-specific trans factors. Two lines of evidence indicate that MyoD (and/or other myfkins) also regulates the promoter of the acetylcholine receptor α subunit via its MEF 1-binding sites. First, wild-type, but not mutant, MEF 1 sites bind to synthetic MyoD in gel shift assays *in vitro* (Piette *et al.*, 1990). Second, mutant promoters cannot be trans activated when co-transfected with MyoD into cultured cells while wild-type promoters are efficiently trans activated in the same assay (Piette *et al.*, 1990). Both these experiments establish a direct correlation between MyoD (or Myfkin) binding and promoter activity, providing compelling evidence that this family of muscle factors, which plays a central role in enhancer function (see above), also plays a role in the regulation of some muscle promoters.

Unexpectedly, a novel line of inquiry indicates that MyoD may also be an important muscle-specific determinant of the human cardiac actin promoter. As discussed above, the muscle-specific activity of the cardiac actin promoter was initially thought to be governed solely by the CArG motif, but this promoter also contains a single MEF 1 motif located between its Sp-1 and TATA motifs (see Fig. 3). The role of the MEF 1 motif was recently revealed by transfection experiments using cultured *Drosophila* Schneider cells, which contain neither Myfkins nor Sp-1 but do contain SRF and other common trans factors such as TFIID (Sartorelli *et al.*, 1990). The human cardiac actin promoter is inactive in transfected schneider cells, demonstrating that SRF and other common factors are not sufficient to activate this promoter. Cotransfection of vectors capable of expressing either MyoD or Sp-1 alone also resulted in the actin promoter remaining inactive. However, when both the MyoD and Sp-1 expression vectors were cotransfected with the cardiac actin promoter full activity was restored. Thus, all three trans factors (SRF, MyoD, and Sp-1) are necessary for full activity of the cardiac actin promoter (Sartorelli *et al.*, 1990). These results leave some questions unanswered: for example, how can CArG elements act as muscle-specific determinants when coupled with heterologous promoters lacking MEF 1 sites? And how does the chicken cardiac actin promoter function without a homologous MEF 1 site? Nevertheless, the results with the human cardiac actin promoter are very exciting because they are the first evidence that muscle gene activity requires collaboration among disparate cis elements and trans factors that were previously thought to work independently.

F. M-CAT Motifs and M-CAT Binding Factor

Another class of muscle regulatory motifs emerged from analysis of the cardiac troponin T gene promoter. As its name implies, the cardiac troponin T gene encodes a cardiac protein. Nevertheless, this gene is also expressed transiently during early skeletal muscle development (Cooper and Ordahl, 1984, 1985; Long and Ordahl, 1988). Analysis of its promoter via transfection into cultured embryonic skeletal muscle demonstrated that determinants of skeletal muscle-specific expression are located within 130 nucleotides of the transcription initiation site while additional upstream elements are required for cardiac-specific expression (Mar *et al.*, 1988, and see below). Chimeric promoter experiments, in which different segments of the cardiac troponin T gene promoter were recombined with corresponding segments of a cell nonspecific promoter, demonstrated that muscle-specific determinants resided in the distal promoter region, between nucleotides -130 and -50 relative to the transcription initiation site at $+1$ (Mar and Ordahl, 1988). The distal promoter region contains sequence motifs that are superficially related to both CArG and MEF 1 (see Fig. 3). Footprint analysis does not indicate binding to the nominal MEF 1 motif but the CArG-like motif is efficiently protected from DNase I digestion by nuclear protein (see Fig. 3 in Mar and Ordahl, 1990). The protein that binds to the CArG-like motif is probably not SRF because competition footprints indicate that the skeletal actin CArG box cannot intercompete for binding. Nevertheless, both of these motifs can be deleted without affecting the muscle-specific activity (Mar and Ordahl, 1990) and thus appear to be superfluous to the activity of the cardiac troponin T gene promoter.

   The deletion experiments outlined above further localized the muscle-specific determinant region to between nucleotides -100 and -50 of the cardiac troponin gene promoter, a region that contains two tandem copies of a novel motif, 5'-CATTCCT-3', dubbed M-CAT for muscle-CAT motif (Mar and Ordahl, 1988, 1990), because identical or related motifs are found in the promoter regions of a number of muscle genes (Nikovits *et al.*, 1986). Mutation of either, or both, copies of the M-CAT motif inactivates the cardiac troponin T gene promoter (Mar and Ordahl, 1988, 1990). Thus, the redundancy of the M-CAT motifs appear to be necessary, like that of the MEF 1 motifs, and unlike that of the CArG motifs (see above). Moreover, even slight alterations in the normal spacing (23 nucleotides, center-to-center) between the two M-CAT motifs substantially reduce the activity of this promoter, suggesting that activity requires an intimate interaction between the factor(s) bound at these two sites (Mar and Ordahl, 1990). The M-CAT motifs differ in this respect from the MEF 1 motifs, whose precise spacing appears to be less important (see Fig. 1).

   The M-CAT motifs are protected from DNase I digestion in footprinting experiments employing nuclear extracts from embryonic muscle (see Fig. 3 for protected regions). In addition, mutations of the M-CAT motif that disrupt promoter function also disrupt the footprint (Mar and Ordahl, 1990). Thus the M-CAT motif is an essential part of the recognition sequence for the binding factor(s) (MCBF for M-CAT binding factor) and that such binding of MCBF is an essential step in the muscle specific activation of transcription from the cardiac troponin T gene promoter.

G. What Is the Significance of Multiple, Muscle-Specific Promoter Motifs?

It is obvious from Fig. 3 that each muscle promoter studied in detail to date has multiple motifs implicated in muscle gene regulation. What is the role, then, of these other motifs? DNA–protein binding experiments show that the M-CAT motif of the skeletal actin promoter binds to MCBF (Mar and Ordahl, 1990) but mutation of this motif has no effect on the activity of the skeletal actin promoter in transfection experiments (Chow and Schwartz, 1990). Similarly, proteins clearly bind to the CArG-like motif of the cardiac troponin T promoter (Mar and Ordahl, 1990) although deletion of this motif has no effect on activity of this promoter in transfection experiments (Mar and Ordahl, 1988, 1990). Only in the case of the cardiac actin promoter is there evidence of collaboration between cis elements and trans factors to achieve muscle-specific activity. Perhaps in the other promoters the apparently superfluous motifs function in ways not evident from transfection of cultured cells, for example, during physiological states not reproduced in culture.

H. Other Elements

Despite the emerging complexity of muscle-specific promoter determinants described above it may be only a glimpse of what is to come. For examples, novel AT-rich motifs (Braun *et al.*, 1989b), an "octa" motif related to those found in immunoglobulin enhancers (Subramaniam *et al.*, 1990), and exon 1 components (Nikovits *et al.*, 1990) have all been implicated as crucial promoter elements in the control of muscle gene transcription. Given this level of complexity with the small number of genes that have been analyzed to date it would seem prudent to expect the diversity of muscle promoter elements to expand considerably in the next few years.

IV. Cardiac Muscle: Distinct Gene Regulatory Components

Although differentiating cardiac muscle cells do not form highly multinucleate myofibers like skeletal muscle, the transformation of the myocardial cytoplasm is comparable to that seen in skeletal muscle cells. Like developing skeletal muscle cells, differentiating cardiac myocytes become densely packed with highly ordered sarcomeric arrays interspersed with the intricate membranous networks of the sarcoplasmic reticular and T-tubular systems. Moreover, the structure, function, and formation of each of these cytoplasmic components in cardiac myocytes are extremely similar to that in skeletal muscle cells. Because the common evolutionary ancestor of cardiac and skeletal muscle cells is extremely ancient, this degree of conservation suggests that very strong selection pressures act on the genetic machinery responsible for striated muscle development.

Paradoxically, despite this extreme level of conservatism, the proteins composing the essential components of the contractile apparatus of a cardiac muscle cell are, for the most part, encoded by a set of genes that is distinct from those encoding the equivalent proteins in skeletal muscle cells. Why two nonoverlapping gene sets are required to produce such similar proteins is somewhat mysterious. On the one hand there is physiological evidence to suggest that each contractile protein isoform may be functionally distinct (albeit if only slightly) from its related cohorts. On the other hand, mutant mice with substantially reduced levels of cardiac actin undergo apparently normal cardiac development and function, indicating that this rigidly conserved protein can be functionally replaced by its skeletal muscle counterpart (Alonso et al., 1990). Thus, the explanation for the rigid conservation and lineage-specific expression of the diversified contractile protein gene families remains problematic.

Equally important, however, are questions regarding the regulation of sarcomeric protein gene transcription in these two cell types. Are the molecular mechanisms governing cardiac-specific transcription the same as those operating in skeletal muscle? The regulation of a few genes has been analyzed in molecular terms in myocardial cells. One of the first such promoters so analyzed was that of the myosin heavy chain; but that analysis concerned itself with the thyroid response element rather than determinants of cardiac-specific expression (Gustafson et al., 1987). The two genes for which the most information is presently available regarding cardiac-specific transcriptional determinants are genes that are also expressed in skeletal muscle. The use of a gene expressed in both types of muscle has the advantage of allowing a direct comparison between the requirements for expression in these two cell types. One such gene is that encoding muscle creatine kinase, which (in mammals) is the creatine kinase isogene expressed in both cardiac and skeletal muscle. Using transgenic mice, Wold and Hauschka and co-workers were able to show that additional upstream sequences are required for expression of the muscle creatine kinase gene in cardiac tissue as compared to that required for skeletal muscle expression (Johnson et al., 1989).

A second gene analyzed in myocardial cells is that encoding cardiac troponin T, which as discussed above has also been analyzed in embryonic skeletal muscle cells, where it is transiently expressed during embryonic development. The analysis of the cardiac troponin T gene was conducted by transfection of embryonic myocardial cells in culture. In this case it was also found that expression in myocardial cells requires additional upstream sequences as compared to those required for skeletal muscle cells (see Fig. 4 and Mar *et al.*, 1988).

Additional information is available for the case of the cardiac troponin T gene (Iannello *et al.*, 1991). The "cardiac element" has been localized to a 47-nucleotide segment of DNA located 200 nucleotides upstream of the transcription initiation site, which, in conjunction with the "minimal" promoter sequences required for skeletal muscle as described above, is necessary for transcription in myocardial cells (see Fig. 4A and B). The cardiac element has some of the properties of an enhancer in that it is capable of acting regardless of orientation or position. An AT-rich region (overlined in Fig. 4E) appears to be important for activity because neither fragment resulting from cleavage at a *Dra*I site within the AT-rich segment (arrow, Fig. 4E) is able to confer cardiac expression alone. Protein binding can be demonstrated to the upstream segment

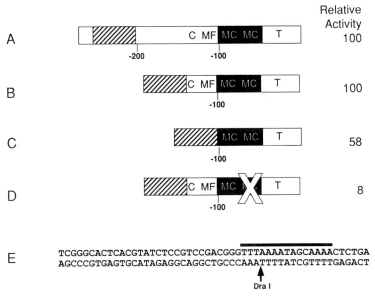

Fig. 4 A cardiac element. (A–D) A series of constructions used to identify a cardiac element (diagonal striped boxes). Numbers indicate nucleotides upstream from the transcription initiation site at +1. Filled box indicates region previously known to contain skeletal muscle-specific determinants (Iannello *et al.*, 1991; Mar and Ordahl, 1988, 1990). C, CArG-like motif; MF, MEF 1 motif; MC, M-CAT motif; T, TATA motif. Large "X" in (D) indicates a site-specific mutation in the proximal M-CAT motif. For details see text and Iannello *et al.* (1991). (E) Nucleotide sequence of the cardiac element. Overlining shows the AT-rich segment with sequence homology to AT-rich segments of muscle enhancers (see Fig. 1).

(Iannello *et al.*, 1991) and recently also to the downstream segment (J. Mar and C. Ordahl, unpublished observations) of the cardiac element, suggesting that its activity is a result of interaction with protein factors in the cell nucleus. The AT-rich segment has a superficial homology to the AT-rich segments of muscle enhancers (compare to Fig. 1) and these two elements may cross-compete in factor binding studies (J. Mar and C. Ordahl, 1991, unpublished observations). Thus, such AT-rich cis elements may represent a point of commonality in gene regulation between cardiac and skeletal muscle.

Are the promoter elements required for expression in myocardial cells the same or different than those required for skeletal muscle expression? This was tested using the cardiac element in conjunction with different portions of the cardiac troponin T promoter (Iannello *et al.*, 1991; also see Fig. 4C and D). The results of those experiments indicated that, similar to the situation found in skeletal muscle cells, activity in myocardial cells requires the M-CAT motifs, but both the CArG-like and MEF 1 motifs are dispensable for activity. Thus, in the cardiac troponin T gene, the factors binding to the cardiac element must collaborate with MCBF in order to activate transcription in myocardial cells. It is not yet known whether this is an absolute requirement for MCBF or if the cardiac element can collaborate with other promoter elements.

The results of the experiments outlined above indicate that the mechanisms governing transcriptional regulation in cardiac myocytes are probably not equivalent to those in skeletal myocytes. In both cases studied to date, additional cis elements are required for expression in cardiac myocytes over and above those required for skeletal muscle expression. This poses intriguing problems for the understanding of the molecular, developmental, and evolutionary relationships between these two closely related cell types.

V. Summary and Perspectives

Analysis of muscle gene transcription has advanced to the point where definitive information is available regarding a number of important muscle-specific cis and trans factors, and information for additional cis and trans factors will soon be forthcoming. An important hurdle, therefore, is the identification and cloning of the relevant trans factors, which will allow detailed analysis of their structure, interactions, and regulation. One class of factors, the Myfkins, is already cloned and progress in understanding the interaction of these factors with their DNA-binding sites and with other transcriptional factors is well under way. It is necessary now to obtain clones for the other binding factors that at present are known only via their DNA-binding sites and footprints. The availability of such clones will not only allow these factors to be analyzed in terms of their binding

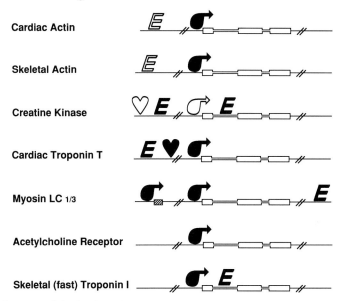

Fig. 5 Summary of the location of muscle-specific determinants in muscle genes. Open boxes indicate exons (wide) and introns (narrow). Thin lines indicate upstream (left) and downstream sequence (right). Arrows indicate promoter regions. Large "E" indicates enhancers. Heart-shaped symbols indicate cardiac elements. Closed symbols indicate regulatory elements known to contain muscle-specific determinants while open symbols indicate regulatory elements not yet determined to contain such determinants. The creatine kinase gene is known to contain two enhancers, one upstream of the promoter and the other within intron I (Jaynes *et al.*, 1988). The myosin light chain 1/3 gene contains two promoters (Donoghue *et al.*, 1988). For other details refer to text.

interactions with DNA and the transcriptional machinery but also in terms of the cooperative interactions between different muscle factors.

The overall picture that seems to be emerging for muscle gene regulation is one of cooperativity between diverse elements. Figure 5 summarizes the location of regulatory regions for a number of different muscle genes. Filled symbols indicate elements that have been shown to contain determinants for muscle-specific transcription. Open symbols represent regulatory regions known to exist but which have not yet been analyzed for the presence of cell-specific determinants. In three cases both the promoter and enhancer regions of the same gene have been examined and in each case both regions have been shown to contain muscle-specific determinants that can function independently from the other (myosin light chain 1/3, cardiac troponin T, and skeletal troponin I; see Fig. 5). First, the skeletal (fast) troponin I gene contains a strong muscle-specific enhancer within its first intron (Konieczny and Emerson, 1987; Yutzey *et al.*, 1989) that is almost certainly Myfkin dependent. The promoter of this gene also contains muscle-specific determinants that have not yet been identified (Nikovits et al., 1990). Second, Myfkins are also likely factors regulating the muscle-specific enhancer of the cardiac troponin T gene (J. Mar and C. Ordahl, 1990, unpublished observations). Myfkins, however, are not required for the muscle-specific promoter of this gene, which depends on M-CAT motifs and the M-CAT binding factor. In this case it is clear that two different trans-acting factor systems can exert effects on the transcriptional machinery either independently or in concert with one another. Finally, as discussed above, the muscle-specific enhancer of the myosin light chain 1/3 gene acts on two promoters, both of which have also been shown to possess as yet unidentified muscle-specific determinants (Billeter et al., 1988).

The above results suggest that the overall muscle specificity of a specific gene results from the synergistic contribution of multiple, independent elements. There is, moreover, a large potential for synergistic interaction of diverse trans-acting factors within individual regulatory domains. The available evidence clearly indicates that Myfkins interact cooperatively with two types of AT-rich binding factors: those that bind to the AT-rich regions of enhancers and those that bind to the AT-rich CArG motif of the cardiac actin promoter. There is potential for other combinatorial interactions. For example, the promoters of the skeletal actin and acetylcholine receptor genes contain M-CAT motifs for which no functional importance is yet evident. Similarly, neither the CArG-like motif nor the MEF 1 motif in the cardiac troponin T gene promoter appear to be required for function. It seems unlikely that such motifs exist without functional significance. In the case of the skeletal actin gene, both the position and sequence of the M-CAT motif have been conserved between birds and mammals, an evolutionary distance too great to be attributed to accident. Thus it is to be expected that functions will emerge for such motifs that are not presently evident in simple transfection experiments. Perhaps such motifs, which appear to be superfluous in transfection experiments, play roles in developmental or physiological processes not reproduced in culture, such as developmental stage-specific repression/activation, or transcriptional responses to physiological changes, If so, the analysis of the interaction among such factors may begin to realize the goal of understanding muscle development and physiology at the level of gene regulation.

Acknowledgments

The author is indebted to many colleagues for helpful discussions and for making unpublished materials available. Generous support for work in the author's laboratory from NIH, NSF, March of Dimes, Muscular Dystrophy Association, and the USDA is also gratefully acknowledged. This review was written during a sabbatical supported by grants from the American Heart Association, the American Cancer Society, and the National Institutes of Health.

References

Alonso, S., Garner, I., Vandekerckhove, J., and Buckingham, M. (1990). Genetic analysis of the interaction between cardiac and skeletal actin gene expression in striated muscle of the mouse. *J. Mol. Biol.* **211,** 727–738.

Bergsma, D., Grichnik, J., Gosset, L., and Schwartz, R. (1986). Delimitization and characterization of cis-acting DNA sequences required for the regulated expression and transcriptional control of the chicken skeletal a-actin gene. *Mol. Cell. Biol.* **6,** 2462–2475.

Billeter, R., Quitschke, W., and Patterson, B. M. (1988). Approximately 1 kilobase of sequence 5' to the two myosin light-chain $1_f/3_f$ gene cap sites is sufficient for differentiation-dependent expression. *Mol. Cell. Biol.* **8**(3), 1361–1365.

Boxer, L. M., Miwa, T., Gustafson, T., and Kedes, L. (1989a). Identification and characterization of a factor that binds to two human sarcomeric actin promoters. *J. Biol. Chem.* **264,** 1284–1292.

Boxer, L. M., Prywes, R., Roeder, R. G., and Kedes, L. (1989b). The sarcomeric actin CArG-binding factor is indistinguishable from the c-fos serum response factor. *Mol. Cell. Biol.* **9**(2), 515–522.

Braun, T., Buschhausen-Denker, G., Bober, E., Tannich, E., and Arnold, H. H. (1989a). A novel human muscle factor related to but distinct from MyoD1 induces myogenic conversion in 10T1/2 fibroblasts. *EMBO J.* **8**(3), 701–709.

Braun, T., Tannich, E., Buschhausen-Denker, G., and Arnold, H.-H. (1989b). Promoter upstream elements of the chicken cardiac myosin light-chain 2A gene interact with trans-acting regulatory factors for muscle-specific transcription. *Mol. Cell. Biol.* **9,** 2513–2525.

Braun, T., Bober, E., Winter, B., Rosenthal, N., and Arnold, H. (1990a). Myf-6, a new member of the human gene family of myogenic determination factors: Evidence for a gene cluster on chromosome 12. *EMBO J.* **9,** 821–831.

Braun, T., Winter, B., Bober, E., and Arnold, H. (1990b). Transcriptional activation domain of the muscle-specific gene-regulatory protein Myf5. *Nature (London)* **346,** 663–665.

Brennan, T., and Olson, E. (1990). Myogenin resides in the nucleus and acquires high affinity for a conserved enhancer element on heterodimerization. *Genes Dev.* **4,** 582–595.

Buskin, J., and Hauschka, S. (1989). Identification of a myocyte nuclear factor that binds to the muscle-specific enhancer of the mouse muscle creatine kinase gene. *Mol. Cell. Biol.* **9,** 2627–2540.

Chow, K. L., and Schwartz, R. J. (1990). A combination of closely associated positive and negative cis-acting promoter elements regulates transcription of the skeletal a-actin gene. *Mol. Cell. Biol.* **10,** 528–538.

Cooper, T., and Ordahl, C. (1984). A cardiac troponin-T gene encodes an embryo-specific isoform in skeletal muscle. *Science* **226,** 979–982.

Cooper, T., and Ordahl, C. (1985). A single troponin T gene generates embryonic and adult isoforms via developmentally regulated alternate splicing. *J. Biol. Chem.* **260,** 11140–11148.

Davis, R. L., Weintraub, H., and Lassar, A. B. (1987). Expression of a single transfected cDNA converts fibroblasts to myoblasts. *Cell (Cambridge, Mass.)* **51,** 987–1000.

Davis, R. L., Cheng, P. F., Lassar, A. B., and Weintraub, H. (1990). The MyoD DNA binding domain contains a recognition code for muscle-specific gene activation. *Cell (Cambridge, Mass.)* **60,** 733–746.

de la Brousse, C., and Emerson, C. (1990). Localized expression of a myogenic regulatory gene, qmf1, in the somite dermatome of avian embryos. *Genes Dev.* **4,** 567–581.

Donoghue, M., Ernst, H., Wentworth, B., Nadal-Ginard, B., and Rosenthal, N. (1988). A muscle-specific enhancer is located at the 3' end of the myosin light-chain ⅓ gene locus. *Genes Dev.* **2**, 1779–1790.

Edmondson, D., and Olson, E. (1989). A gene with homology to the myc similarity region of MyoD1 is expressed during myogenesis and is sufficient to activate the muscle differentiation program. *Genes Dev.* **3**, 628–640.

Gossett, L. A., Kelvin, D. J., Sternberg, E. A., and Olson, E. N. (1989). A new myocyte-specific enhancer-binding factor that recognizes a conserved element associated with multiple muscle-specific genes. *Mol. Cell. Biol.* **9**, 5022–5033.

Grichnik, J., Bergsma, D., and Schwartz, R. (1986). Tissue restricted and stage specific transcription is maintained within 411 nucleotides flanking the 5' end of the chicken a-skeletal actin gene. *Nucleic Acids Res.* **14**, 1683–1701.

Gustafson, T., Markham, B., Bahl, J., and Morkin, E. (1987). Thyroid hormone regulates expression of a transfected a-myosin heavy chain fusion gene in fetal heart cells. *Proc. Natl. Acad. Sci. U.S.A.* **84**, 3122–3126.

Hopwood, N. D., Pluck, A., and Gurdon, J. B. (1989). MyoD expression in the forming somites is an early response to mesoderm induction in *Xenopus* embryos. *EMBO J.* **8**(11), 3409–3417.

Horlick, R. A., and Benfield, P. A. (1989). The upstream muscle-specific enhancer of the rat muscle creatine kinase gene is composed of multiple elements. *Mol. Cell. Biol.* **9**, 2396–2413.

Horlick, R. A., Hobson, G. M., Patterson, J. H., Mitchell, M. T., and Benfield, P. A. (1990). Brain and muscle creatine kinase genes contain common TA-rich recognition protein-binding regulatory elements. *Mol. Cell. Biol.* **10**, 4826–4836.

Iannello, R. C., Mar, J. H., and Ordahl, C. P. (1991). Characterization of a promoter element required for transcription in myocardial cells. *J. Biol. Chem.* **266**, 3309–3316.

Jaynes, J. B., Johnson, J. E., Buskin, J. N., Gartside, C. L., and Huschka, S. D. (1988). The muscle creatine kinase gene is regulated by multiple upstream elements, including a muscle-specific enhancer. *Mol. Cell. Biol.* **8**, 62–70.

Johnson, J., Wold, B., and Hauschka, S. (1989). Muscle creatine kinase sequence elements regulating skeletal and cardiac muscle expression in transgenic mice. *Mol. Cell. Biol.* **9**, 3393–3399.

Kodonaga, J., Jones, K., and Tjian, R. (1986). Promoter specific activation of RNA polymerase II transcription by Sp 1. *Trends Biochem.* **11**, 20–23.

Konieczny, S., and Emerson, C. (1987). Complex regulation of the muscle-specific contractile protein (troponin I) gene. *Mol. Cell. Biol.* **7**, 3065–3075.

Lassar, A. B., Buskin, J. N., Lockshon, D., Davis, R. L., Apone, S., Hauschka, S. D., and Weintraub, H. (1989). MyoD is a sequence-specific DNA binding protein requiring a region of myc homology to bind to the muscle creatine kinase enhancer. *Cell (Cambridge, Mass.)* **58**, 823–831.

Lassar, A. B., Davis, R. L., Wright, W. E., Kadesch, T., Murre, C., Voronova, A., Baltimore, D., and Weintraub, H. (1991). Functional activity of myogenic HLH proteins requires heterooligomerization with E12/E47-like proteins *in vivo*. *Cell (Cambridge, Mass.)* **66**, 305–316.

Lin, A., Dechesne, C., Eldridge, J., and Paterson, B. (1989). An avian muscle factor related to MyoD1 activates muscle-specific promoters in nonmuscle cells of different germlayer origin and in BrdU-treated myoblasts. *Genes Dev.* **3**, 986–996.

Long, C., and Ordahl, C. (1988). Transcriptional repression of an embryo-specific muscle gene. *Dev. Biol.* **127**, 228–234.

Mar, J., and Ordahl, C. (1988). A conserved CATTCCT motif is required for skeletal muscle specific expression of the cardiac troponin T gene promoter. *Proc. Natl. Acad. Sci. U.S.A.* **85**, 6404–6408.

Mar, J., and Ordahl, C. P. (1990). M-CAT binding factor, a novel trans-acting factor governing muscle-specific transcription. *Mol. Cell. Biol.* **10**, 4271–4283.

Mar, J., Antin, P., Cooper, T., and Ordahl, C. (1988). Analysis of the upstream regions governing expression of the chicken cardiac troponin T gene in embryonic cardiac and skeletal muscle cells. *J. Cell Biol.* **107**, 573–585.

Miner, J., and Wold, B. (1990). Herculin, a fourth member of the MyoD family of myogenic regulatory genes. *Proc. Natl. Acad. Sci. U.S.A.* **87**, 1089–1093.

Minty, A., and Kedes, L. (1986). Upstream regions of the human cardiac actin gene that modulate its transcription in muscle cells: Presence of an evolutionarily conserved repeated motif. *Mol. Cell. Biol.* **6**, 2125–2136.

Miwa, T., Boxer, L., and Kedes, L. (1987). CArG boxes in the human cardiac a-actin gene are core binding sites for positive trans-acting regulatory factors. *Proc. Natl. Acad. Sci. U.S.A.* **84**, 6702–6706.

Mohun, T. J., Taylor, M. V., Garrett, N., and Gurdon, J. B. (1989). The CArG promoter sequence is necessary for muscle-specific transcription of the cardiac actin gene in *Xenopus* embryos. *EMBO J.* **8**, 1153–1161.

Nikovits, W., Kuncio, G., and Ordahl, C. P. (1986). The chicken fast skeletal troponin I gene: Exon organization and sequence. *Nucleic Acids Res.* **8**, 3377–3390.

Nikovits, W., Mar, J., and Ordahl, C. (1990). Muscle-specific activity of the skeletal troponin I promoter requires interaction between upstream regulatory sequences and elements contained within the 1st transcribed exon. *Mol. Cell. Biol.* **10**, 3468–3482.

Nudel, U., Greenberg, D., Ordahl, C., Saxel, O., Neuman, S., and Yaffe, D. (1985). Developmentally regulated expression of a chick muscle-specific gene in stably transfected rat myogenic cell. *Proc. Natl. Acad. Sci. U.S.A.* **82**, 3106–3109.

Olson, E. N. (1990). MyoD family: A paradigm for development? *Genes Dev.* **4**, 1454–1461.

Ordahl, C., and Cooper, T. A. (1983). Strong homology in the promoter and 3' untranslated regions of the chick and rat alpha-actin gene. *Nature (London)* **303**, 348–349.

Piette, J., Klarsfeld, A., and Changeux, J.-P. (1989). Interaction of nuclear factors with the upstream region of the a-subunit gene of chicken muscle acetylcholine receptor: Variations with muscle differentiation and denervation. *EMBO J.* **8**, 687–694.

Piette, J., Bessereau, J.-L., Huchet, M., and Changeux, J.-P. (1990). Two adjacent MyoD1-binding sites regulate expression of the acetylcholine receptor a-subunit gene. *Nature (London)* **345**, 353–355.

Pinney, D., Pearson-White, S., Konieczny, S., Latham, K., and Emerson, C. (1988). "Myogenic lineage determination and differentiation: Evidence for a regulatory gene pathway." *Cell (Cambridge, Mass.)* **53**, 781–793.

Rhodes, S., and Konieczny, S. (1989). Identification of MRF4: A new member of the muscle regulatory gene family. *Genes Dev.* **3**, 2050–2061.

Sartorelli, V., Webster, K. A., and Kedes, L. (1990). Muscle-specific expression of the cardiac a-actin gene requires MyoD1, CArG-box binding factor, and Sp1. *Genes Dev.* **4**, 1811–1822.

Sassoon, D., Lyons, G., Wright, W., Lin, V., Lassar, A., Weintraub, H., and Buckingham, M. (1989). Expression of two myogenic regulatory factors myogenin and MyoD1 during mouse embryogenesis. *Nature (London)* **341**, 303–307.

Scales, J., Olson, E., and Perry, W. (1990). Two distinct *Xenopus* genes with homology to MyoD1 are expressed before somite formation in early embryogenesis. *Mol. Cell. Biol.* **10**, 1516–1524.

Sternberg, E., Spizz, G., Perry, M., and Olson, E. (1989). A ras-dependent pathway abolishes activity of a muscle-specific enhancer upstream from the muscle creatine kinase gene. *Mol. Cell. Biol.* **9**, 594–601.

Subramaniam, A., Gulick, J., and Robbins, J. (1990). Analysis of the upstream regulatory region of a chicken skeletal myosin heavy chain gene. *J. Biol. Chem.* **265**, 13986–13994.

Tapscott, S. J., Davis, R. L., Thayer, M. J., Cheng, P. F., Weintraub, H., and Lassar, A. B. (1988). MyoD1: A nuclear phosphoprotein requiring a myc homology region to convert fibroblasts to myoblasts. *Science* **242**, 405–411.

Charles P. Ordahl

Thompson, W. R., Nadal-Ginard, B., and Mahdavi, V. (1991). A MyoD1-independent muscle-specific enhancer controls expression of the β-myosin heavy chain gene in skeletal and cardiac muscle cells. *J. Biol. Chem.* **266,** in press.

Treisman, R. (1986). Identification of a protein-binding site that mediates transcriptional response of the c-fos gene to serum factors. *Cell (Cambridge, Mass.)* **46,** 567–574.

Tuil, D., Clergue, N., Montarras, D., Pinset, C., Kahn, A., and Phan-Dinh-Tuy, F. (1990). CC Ar GG boxes, cis-actin elements with dual specificity. *J. Mol. Biol.* **213,** 677–686.

Uetsuki, T., Nabeshima, Y., Fujisawa-Sehara, A., and Nabeshima, Y.-I. (1990). Regulation of the chicken embryonic myosin light-chain (L23) gene: Existence of a common regulatory element shared by myosin alkali light-chain genes. *Mol. Cell. Biol.* **10,** 2562–2569.

Walsh, K. (1989). Cross-binding of factors to functionally different promoter elements in c-*fos* and skeletal actin genes. *Mol. Cell. Biol.* **9**(5), 2191–2201.

Walsh, K., and Schimmel, P. (1987). Two nuclear factors compete for the skeletal muscle actin promoter. *J. Biol. Chem.* **262,** 9429–9432.

Wang, X.-M., Tsay, J.-J., and Schmidt, J. (1990). Expression of the acetylcholine receptor d-subunit gene in differentiating chick muscle cells is activated by an element that contains two 16 bp copies of a segment of the a-subunit enhancer. *EMBO J.* **9,** 783–790.

Wang, Y., Xu, H. P., Wang, X. M., Ballivet, M., and Schmidt, J. (1988). A cell type-specific enhancer drives expression of the chick muscle acetylcholine receptor a-subunit gene. *Neuron* **1,** 527–534.

Weintraub, H., Davis, R. L., Lockshon, D., and Lassar, A. B. (1990). MyoD binds cooperatively to two sites in a target enhancer sequence: Occupancy of two sites is required for activation. *Proc. Natl. Acad. Sci. U.S.A.* **87,** 5623–5627.

Wentworth, B., Donoghue, M., Engert, J., Berglund, E., and Rosenthal, N. (1991). Paired MyoD binding sites regulate myosin light gene expression. *Proc. Natl. Acad. Sci. U.S.A.* **88,** 1242–1246.

Wright, W. E., Sassoon, D. A., and Lin, V. K. (1989). Myogenin: A factor regulating myogenesis, has a domain homologous to MyoD. *Cell (Cambridge, Mass.)* **56,** 607–617.

Yutzey, K., Kline, R., and Konieczny, S. (1989). An internal regulatory element control troponin I gene expression. *Mol. Cell. Biol.* **9,** 1397–1405.

Index

A

Acetylcholine receptor, sarcomeric gene expression and, 152–153, 156–157, 164

Acetylcholinesterase, sarcomeric gene expression and, 152

Actin, *see also* F-actin; G-actin
  bicoid message localization and, 28
  brush border cytoskeleton and, 94–98, 112–113, 115
  differentiation, 109–110
  embryogenesis, 102, 104–107
  cardiac, 153, 155–157, 160, 164
  chicken intestinal epithelium and, 126, 129, 131–135, 137
  cytoskeleton positional information and, 2–3
  gastrulation and, 4
  organization in sea urchin egg cortex, 9–10, 18–19
    actin-binding proteins, 16–18
    assembly states, 11–14
    fertilization cone, 14
    organization changes, 14–16
  sarcomeric gene expression and, 153, 155–160, 164
  skeletal, 153, 155–156, 158–159, 164
  tissue-specific cytoskeletal structures and, 4–5

Actin-binding proteins
  actin organization in sea urchin egg cortex and, 16–18
  brush border cytoskeleton and, 94, 104, 107, 112
  chicken intestinal epithelium and, 129
  in *Xenopus*, 35, 50–51
    fertilization, 37–38
    gastrulation, 38–41
    MAb 2E4 antigen, 41–50
    oogenesis, 36–37

Actin-bundling proteins, chicken intestinal epithelium and, 129, 135–136

α-Actinin
  brush border cytoskeleton and, 109
  chicken intestinal epithelium and, 129, 133

Actomyosin, chicken intestinal epithelium and, 126

Adenocarcinomas
  brush border cytoskeleton and, 113–114
  chicken intestinal epithelium and, 135

Adhesion, *Xenopus* actin-binding proteins and, 40

ADP, actin organization in sea urchin egg cortex and, 11

Amino acids
  microtubule motors in sea urchin and, 80
  sarcomeric gene expression and, 151

Annexins, chicken intestinal epithelium and, 136–138

Antibodies
  actin organization in sea urchin egg cortex and, 12, 17
  brush border cytoskeleton and, 107–108
  microtubule motors in sea urchin and, 76–77, 79
  reorganization in frog egg and, 59–60, 67
  Xenopus actin-binding proteins and, 37, 41, 44, 48, 50

Antigens, *Xenopus* actin-binding proteins and, 41–50

Asters, reorganization in frog egg and, 58, 62

Asymmetry
  bicoid message localization and, 23, 27
  cytoskeleton positional information and, 1, 3
  reorganization in frog egg and, 53–54, 60, 62, 64
  Xenopus actin-binding proteins and, 38–39

ATP
  microtubule motors in sea urchin and, 71, 76–77, 79–80, 82
  Xenopus actin-binding proteins and, 38, 41

ATPase
  actin organization in sea urchin egg cortex and, 11
  microtubule motors in sea urchin and, 75–77, 79–80, 82, 86